MECHANICAL ENGINEERING THEORY AND APPLICATIONS

NOVIKOV GEARING: ACHIEVEMENTS AND DEVELOPMENT

MECHANICAL ENGINEERING THEORY AND APPLICATIONS

Additional books in this series can be found on Nova's website under the Series tab.

Additional E-books in this series can be found on Nova's website under the E-books tab.

MECHANICAL ENGINEERING THEORY AND APPLICATIONS

NOVIKOV GEARING: ACHIEVEMENTS AND DEVELOPMENT

VIKTOR I. KOROTKIN
NIKOLAY P. ONISHKOV
AND
YURI D. KHARITONOV

Nova Science Publishers, Inc.
New York

LIBRARY OF CONGRESS CATALOGING-IN-PUBLICATION DATA

Korotkin, Viktor I.
Novikov gearing : achievements and development / Viktor I. Korotkin, Nikolay P. Onishkov, Yury D. Kharitonov.
p. cm.
Includes bibliographical references and index.
ISBN 978-1-61761-193-3 (hardcover)
1. Gearing, Novikov. I. Onishkov, Nikolay P. II. Title.
TJ202.K668 2010
621.8'33--dc22
2010029792

Published by Nova Science Publishers, Inc. † New York

CONTENTS

FOREWORD

Since the Novikov gearing appeared in the middle of the 1950s, a considerable number of theoretical and experimental investigations has been carried out and a mass-scale implementation of this gearing into several branches of industry has been performed.

It should be noted that rather quick and effective implementation of Novikov gearing took place in cases when the hardness of tooth flanks was under *HB 350*, that is, when the limiting factor of the gearing working capacity was a contact endurance according to which Novikov gearing possesses undeniable advantages over the involute one.

As for the gearing with high-hardened teeth, the situation is different here – at first the attempts to accelerate its implementation led to frequent teeth failure, and that prevented the progressive high-hardened Novikov gearing from industrial application. It was explained by a deficiency of sufficient grounding of the basic rack profile parameters and geometry as a whole, perfect design methods and a necessary number of tests.

Some time later, scientists paid their attention to Novikov gearing with hardened teeth due to sharply increased requirements to the loading level of power drives and the development of manufacture technique and design methods of this gearing. The first large implementations appeared. Soon, the problem of application of Novikov gearing with high-hardened teeth appeared and became so timely that in 1980, a special Coordination Board was created at the USSR Minstankoprom on implementation of such gearing in general-purpose gearbox industry. Under its guidance, large-scale theoretical and experimental researches were carried out that led to the implementation of Novikov gearing at Izhevsk "OOO "Reduktor".

The monograph contains a review of a problem of Novikov gearing practical application and brief systematization of results of long-term tests of this gearing in different organizations. Characteristics of now- widespread basic rack profiles, including the basic rack profile according to the State standard GOST 30224-96, developed with the direct assistance of the authors and standardized within the framework of Commonwealth of Independent States (CIS) for the gearing with hardened tooth flanks, are described, and engineering techniques are proposed for geometrical and strength analysis of Novikov gearing, approved by Russia State Standard Organization and intended for standardization.

The monograph content is based on the research of the authors with a wide application of theoretical and experimental results obtained by other researchers.

All calculations, conclusions and recommendations refer to Novikov gearing with two lines of action (NG TLA), which is the most widely used.

The strength analyses described in the monograph are based on modern achievements of the analytical theory of elasticity and its numerical methods The final form of the proposed relations makes it possible to give an engineering evaluation of Novikov gearing strength rather quickly and with acceptable accuracy, that was verified by wide-range numerical simulation and comparison with results of nature tests.

The expressions for calculation of acting stresses (left side of strength conditions) were developed, taking into account the possibility of accepting median stress values, described in the standard GOST 21354-87 for involute gearing, as the limiting values (which are components of the right side of strength conditions). Here, the choice of the safety factor, and, therefore, the allowable stresses, is particularly specified. The SI system is used as a system of values measurement. At the head of the monograph, the basic letter designations and definitions are given, which are repeatedly met in the text.

The represented monograph is a logical continuation and development of two previous ones: Cylindrical Novikov gearing / Korotkin V.I., Roslivker E.G., Pavlenko A.V., Veretennikov V.Ya., Kharitonov Yu.D. Kiev: Tekhnika, 1991 152 p. and Korotkin V.I., Kharitonov Yu.D. Novikov gearing. Rostov-on-Don: RSU Publishers. 1991. 208 p.

The present monograph is complemented with the results of a large number of research works carried out by the authors within the period from 1991 to 2006 and published in central and local press. These research works are concerned with non-uniform distribution of the transmitted load, bending and contact stresses among contact areas, internal dynamics of engagement, contact friction, contact stresses when the contact area shifts to the tooth edge (so-called “boundary effect”), gearing adaptability as the most important, but insufficiently studied feature of Novikov gearing, and reasonable choice of the design safety factors. The monograph contains information about new results of depth -contact strength investigation for surface strengthened teeth as one of the failure criteria of the latter, problems of manufacture techniques of bevel Novikov gearing with the assigned properties (gearing synthesis) and peculiarities of its strength analysis. A description and principles of formation of original types of cylindrical Novikov gearing, developed with the assistance of the authors, which, for example, don’t have (or almost don’t have) the axial force component in the engagement, are given, that widens the area of Novikov gearing application. The important section is a description of quite effective, and at the same time technologically simple, methods of considerable increase (sometimes up to two times and more) of the load-bearing capacity of cylindrical and bevel Novikov gearing, based on equalization of stresses over contact areas in multipoint engagement.

Application of the monograph information allows to design and calculate Novikov gearing more reasonably and correctly than it was before, taking into account real conditions of its operation, to define its optimal geometrical characteristics, thus, increasing the load-bearing capacity, reducing the material consumption and cost price of production, and, finally, making the gear drive more competitive in the world market.

Therefore, the monograph purpose is to feasibly assist the employees of enterprises and design departments to design and manufacture Novikov gearing reasonably, and also to acquaint researchers, teaching instructors, post-graduates and students of higher educational institutions with the latest achievements in the given area.

Chapters 7 and 17 (except for the paragraph 17.7) and Appendix 5 were written by Ph.D. Kharitonov Yu.D.; Chapter 12 and Example 2 in Appendix 3 were written by Ph.D.

Onishkov N.P. The rest of the monograph material, including Appendices, was written by Ph.D. Korotkin V.I.

The Laboratory of Special Gearing of Vorovich Mechanics and Applied Mathematics Research Institute, Southern Federal University is ready to give assistance to the interested organizations in carrying out the optimal design and calculation of cylindrical and bevel Novikov gearing on the basis of the available software package, and also to share these packages on the contract basis.

Reviews of the monograph, and requests on purchasing the software for geometrical and strength analysis of cylindrical and bevel Novikov gearing should be sent to the address: Russia, 344090, Rostov-on-Don, Prospect Stachki, 200/1, VMAMRI SFU, Laboratory of Special Gearing, for Korotkin V.I. Phones: 8-(863)2975223, 8-(863)2227152, e-mail: korotkin@math.rsu.ru

BASIC LETTER DESIGNATIONS AND DEFINITIONS

Letter designations, common to both gears of the engaging pair, with subscripts "*1*" refer to the driving pinion, with subscript "*2*" – to the driven gear, without the noted indexes they refer equally to both elements of the pair.

The term "gear" refers both to the pinion and the gear.

The asterisk in the upper part of the letter designation indicates that the given linear value corresponds to the unit normal tooth module.

DESIGNATIONS REFERRING TO THE BASIC RACK PROFILE

ρ_a - radius of arc of the addendum active part;

ρ_f - radius of arc of the dedendum active part;

$\Delta\rho$ - difference of radii ρ_f - ρ_a;

$\rho_a^{'}$ - radius of arc of the additional convex part of the addendum;

ρ_p - radius of arc of the concave conversion area;

ρ_q - radius of arc of conjugation with arcs of the active dedendum and conversion area;

ρ_i - radius of arc of the dedendum fillet;

x_a, l_a - distance from the arc center with radius ρ_a to the pitch line and tooth symmetry axis, correspondingly;

$x_a^{'}$, $l_a^{'}$ - distance from the arc center with radius $\rho_a^{'}$ to the pitch line and tooth symmetry axis, correspondingly;

x_f, l_f - distance from the arc center with radius ρ_f to the pitch line and tooth symmetry axis, correspondingly;

x_p, l_p - distance from the arc center with radius ρ_p to the pitch line and tooth symmetry axis, correspondingly;

x_q, l_q - distance from the arc center with radius ρ_q to the pitch line and tooth symmetry axis, correspondingly;

h_a, h_f - height of the pitch addendum and dedendum, correspondingly;

h - total tooth height;

t_a, t_f - distance from the pitch line to the beginning of the active part of addendum and dedendum profiles, correspondingly;

l - value of the arc of the addendum active part measured along the chord;

α_k - tooth profile angle at the nominal (theoretical) contact point;

α_a, α_p - maximum and minimum profile angles of the addendum active part, correspondingly;

α_{p1}, α_{p2} - maximum and minimum profile angles of the additional convex part of the addendum, correspondingly;

α_f - minimum profile angle of the dedendum active part;

α - inclination angle of the straight-line conversion part to the axis of tooth symmetry;

j_k - nominal backlash in the pair of conjugate basic gear racks, provided by the basic rack profile configuration.

Subscripts "*a*" and "*f*" refer to the addendum and dedendum, correspondingly.

Designations Referring to Geometry of Cylindrical Gearing

m - normal tooth module;

z - number of gear teeth;

z_v - reduced tooth number of the equivalent straight-tooth gear;

a_w - center distance in the engagement of the gear pair;

a - pitch center distance of the gear pair;

x - shift of the rack when cutting the gear;

x_Σ - total shift of racks when cutting gears of the pair;

α_{ktw} - face pressure angle of the gear pair;

α_{kw} - normal pressure angle of the gear pair;

β - helix angle of the rack (or helix angle of the gear at the pitch cylinder);

β_w - helix angle of the gear at the reference cylinder;

b - width of toothing of the gear;

b_w - operating width of toothing of the gear pair;

$d\ (r)$ - pitch diameter (radius) of gear;

$d_w(r_w)$ - reference diameter (radius) of gear;

d_a, d_f - tip and root diameter of gear teeth, correspondingly;

p_x - axial pitch of the rack teeth;

ε_β - axial overlap ratio of teeth;

ε_q - phase axial overlap ratio of teeth;

u - gear ratio of the pair, equal to z_2/z_1 ;

ψ_{ba} - ratio b_w/a_w;

ψ_{bd} - ratio b_w / d_1;

j_w - backlash in engagement of the gear pair;

x_{w1}, x_{w2} - distance from the gear pitch point line to the pitch surface of the pinion and gear, correspondingly;

x_w - distance from the gear pitch point line to the pitch plane of the basic rack;

ϑ_n - profile angle of the basic rack addendum, corresponding to undercut of the gear addendum profile;

S_a - normal tooth thickness of the gear at the tip surface;

l_z - value of the active part of the gear addendum (dedendum);

K_l - ratio l_z / l;

ρ_α - profile reduced main curvature radius of contacting flanks at the theoretical contact point;

ρ_β - longitudinal reduced main curvature radius of contacting flanks at the theoretical contact point;

$C_{\alpha\beta}$ - ratio ρ_α / ρ_β.

Designations Referring to Accuracy, Control and Adaptability of Cylindrical Gearing

$F_{\beta r}(F_\beta)$ - tooth direction error (direction tolerance);

$f_{fr}(f_f)$ - tooth profile error (profile error tolerance);

$f_{ptr}(f_{pt})$ - tooth circular pitch error (tolerance for difference of pitches);

$f_{xr}(f_x)$ - axes parallelism error (parallelism error tolerance) of gears;

$f_{yr}(f_y)$ - axes misalignment (misalignment tolerance) of gears;

$F_{rr}(F_r)$ - radial run-out (radial run-out tolerance) of the gear rim;

$f_{ar}(f_a)$ - center distance error (limiting deviation) of the gear pair;

h_x - measuring tooth height to the chord;

S_x - measuring tooth thickness along the chord;

z_n - tooth number of the gear, gripped by the snap gauge when determining the common normal length;

W - measuring length of the common normal;

W_y - common normal length along the prominent part of the dedendum;

Δr - manufacturing reduced radial error of the gear pair;

Δa - radial error taken by the gearing;

t_c - percent of the “risk level” in probability analysis;

$\widetilde{T}_H$ - tolerance for the shift of the basic rack profile when cutting the gear teeth;

$\widetilde{T}_{H\Sigma}$ - total (for the pinion and gear) tolerance for the shift of the basic rack profile when tooth-cutting;

$\widetilde{T}_{\underline{H}}$, $\widetilde{T}_{\overline{H}}$ - lower and upper boundary of the limiting deviation of the basic rack profile shift when cutting the gear teeth, correspondingly;

$\widetilde{T}_S$ - tolerance for the measuring tooth thickness along the chord;

$\widetilde{T}_{\underline{S}}$, $\widetilde{T}_{\overline{S}}$ - lower and upper boundary of the limiting deviation of the measuring tooth thickness along the chord, correspondingly;

$\widetilde{T}_W$ - tolerance for the measuring length of the common normal;

$\widetilde{T}_{\underline{W}}$, $\widetilde{T}_{\overline{W}}$ - lower and upper boundary of the limiting deviation of the measuring length of the common normal, correspondingly;

k - number of the degree of kinematic accuracy or number of the reduced degree (according to smoothness and contact ratings) of accuracy of the gear;

k_p - number of the degree of gear accuracy according to smoothness rating;

k_k - number of the degree of gear accuracy according to contact rating.

Designations Referring to Strength Analysis of Cylindrical Gearing

H - material hardness (symbol);

$H_{HB}(HB)$ - Brinell hardness of the material;

H_{HRC} - Rockwell hardness of the material;

$H_{HV}(HV)$ - Vickers hardness of the material;

H_0, H_e, H_k - maximum hardness of the surface area, effective hardness and core hardness, correspondingly;

H^z - hardness at the design depth z;

F_n - normal force in the engagement;

F_t - tangential force in the engagement;

F_{rad} - radial force in the engagement;

F_x - axial force in the engagement;

T - torque transmitted by the gearing;

T_H - design torque in the contact endurance analysis;

T_F - design torque in the bending endurance analysis;

σ - stress (symbol);

σ_H - normal contact stress;

σ_F - bending stress;

$\sigma_{k\,max}$ - maximum normal contact stress at the center of the contact ellipse;

σ_{HK} - normal contact depth stress;

σ_e - effective (equivalent, reduced) stress (symbol);

σ_{HKe} - effective contact depth stress;

σ^0_{HKe} - standard contact depth stress, equal to $\sigma_{HKe} / \sigma_{k\,max}$;

σ_{HKP} - allowable normal contact depth stress;

σ_{HKP_e} - allowable effective contact depth stress;

σ^0_{HKPe} - standard allowable effective contact depth stress, equal to $\sigma_{HKPe}/\sigma_{k\,max}$;

σ_{HKlime} - limit of the contact depth endurance according to effective stresses;

$\sigma_1, \sigma_2, \sigma_3$ - main stresses;

σ_i - intensity of octahedral stresses;

$\sigma_{vr+}, \sigma_{vr-}$ - material ultimate strength at tension and compression, correspondingly;

$\sigma_{0,2+}, \sigma_{0,2-}$ - conventional yield point of material under tension and compression, correspondingly;

σ_+, σ_- - stress of endurance breakage (failure) of material under single-axis tension and compression, correspondingly;

$\chi = \sigma_+ / \sigma_-$ - parameter of material plasticity;

τ_{max} - maximum tangential stress;

τ_{yz} - orthogonal tangential stress;

N - number of cycles of stress alternation (symbol);

N_E - equivalent number of cycles of stress alternation;

N_{HKlim}, N_{EK} - basic and equivalent number of cycles of stress alternation, correspondingly, in depth contact strength analysis;

E - elasticity modulus;

ν - Poisson's ratio;

$S_H (S_F)$ - safety factor in surface contact (bending) endurance analysis;

S_{HK} - safety factor in depth contact endurance analysis;

K_H - load factor in contact strength analysis;

K_F - load factor in bending strength analysis;

K_T - coefficient taking into account the non-uniform distribution of the transmitted load among contact areas;

K_b - coefficient taking into account the influence of the gear face on tooth displacement and bending stresses;

K^0_σ - coefficient of non-uniform stress distribution among contact areas (stress concentration factor) before the run-in of teeth (initial);

K_σ - stress concentration factor after the run-in of teeth;

K_A - coefficient taking into account the external dynamic load;

K_λ - coefficient taking into account the gearing adaptability;

K_f - coefficient taking into account the influence of friction forces in contact;

K_c - coefficient of stress distribution among contact areas for the gearing, manufactured without errors and having an absolutely rigid structural design;

K_w - coefficient taking into account the run-in of tooth flanks;

K_v - coefficient taking into account the dynamic component of the force in the engagement;

k_e - equivalency factor;

a_H - major semi-axis of the contact ellipse;

b_H - minor semi-axis of the contact ellipse;

$\overline{\beta}$ - ellipticity factor, equal to b_H / a_H;

v - tangential speed of gears;

Z_l - coefficient of run-in completeness of tooth flanks;

Y_v - tooth shape coefficient;

Y_a - coefficient taking into account the longitudinal expansion of the contact area along the tooth length;

Z_N - durability factor in surface contact endurance analysis;

Y_N - durability factor in bending endurance analysis;

Z_{LK} - durability factor in depth contact strength analysis;

h_0 - distance from the surface to the layer with hardness $H = H_0$;

h_{te}, h_t - effective and total thickness of the hardened layer, correspondingly;

z^0 - relative depth of the layer, equal to z/b_H;

W_δ - total elastic tooth displacement in the contact point;

W_k - elastic tooth displacement caused by the compliance of mating parts;

δf - longitudinal tooth flank depth;

δl - ease-off of the barrel-shaped tooth surface at the gear face.

Loads, stresses and a number of other parameters have several subscripts: "*H*" relates to the analysis of surface contact endurance, "*HK*" - of depth contact endurance, "*F*" - of bending endurance, "*max*" - of strength under the action of maximum load; index "*e*" means the effective value of stress, index "*P*" marks allowable and index "*lim*" - limiting stresses, loads, etc.; the superscript "*E*" relates to parameters obtained experimentally; the superscript "*z*" relates to parameters determined at the depth "*z*" of the diffusion layer; for example, σ^z_{HKPe} - allowable effective contact depth stress at the depth "*z*", T_{FP} - allowable torque in bending endurance analysis, etc.

SOME ACCEPTED ABBREVIATIONS:

CHT - chemically heat treatment;
C - carburizing;
NC - nitro carburizing;
NG – Novikov gearing;

OLA – one line of action;
TLA – two lines of action;
NG OLA – Novikov gearing with one line of action;
NG TLA – Novikov gearing with two lines of action;
SSS – stress-strain state;
DCS – depth contact strength;
DCF - depth contact failure.

Designations Referring to Bevel Gearing

The following subscripts refer: "*n*" - to parameters describing the section, normal to the tooth direction, "*e*" - to parameters describing the external face of the gear, "*i*" - to parameters describing the internal face of the gear, "*0*" - to parameters describing the gear-cutting tool, "*c*" - to parameters describing the plane generating gear.

m_0 - tool module;

m_n - mean normal module of teeth;

R - mean cone distance;

R_z - current cone distance;

x_n^* - mean normal shift coefficient;

x_τ^* - coefficient of the design tooth thickness alternation;

z_{vn} - tooth number of the biequivalent cylindrical gear;

z_c - tooth number of the generating gear;

δ - pitch cone angle of the gear;

β_n - mean normal design helix angle of the tooth;

δl_T - tooth ease-off in the face plane at longitudinal modification;

δW_δ - total angular elastic displacement of teeth at the contact point;

δW_k - angular elastic tooth displacement caused by the compliance of mating parts;

q - run-in parameter (angular setting of the generating surface);

U - radial setting of the generating surface;

i_c - ratio of angular speeds of the cut and generating gears;

a_w - shift of the pitch cone apex relative to the generating gear axis in direction, perpendicular to axes of the gear pair;

ΔR - shift of the pitch cone apex relative to the generating gear axis in direction along the generatrix of the gear pitch cone.

Abbreviations of Organizations, Enterprises and Institutes

VMAMRI – Vorovich Mechanics and Applied Mathematics Research Institute;
SFU – Southern Federal University;
RSU – Rostov State University;

ZAFEA – Zhukovsky Air-Force Engineering Academy;
LMI – Leningrad Mechanical Institute;
"OOO "Reduktor" – Limited Liability Company "Reduktor" (Izhevsk, Russia);
LSG - Laboratory of Special Gearing.

INTRODUCTION

One of the main purposes of gearing synthesis is the reduction of its material consumption.

Overall dimensions and material consumption of involute gearing dominating in engineering industry are determined by contact endurance of active flanks. There had been a lot of attempts to develop systems of engagement exceeding the involute one in contact endurance, among which the investigation results of Yu.N. Budyka [13] and E. Wildhaber [259] should be marked out. Nevertheless, these attempts did not lead to the required result until fundamental developments made by M.L. Novikov [158].

It is known that the dangerous area in involute gearing according to the contact endurance is the area located at the dedendum near the pitch point line, although the reduced tooth profile curvature in this area is less than at the pinion tooth root. It is explained, firstly, by unfavorable direction of fatigue cracks at which lubrication is pressed deep inside [221], secondly, by the increased friction factor near the pitch point line, resulting in the rise of effective contact stresses [95, 96] and, thirdly, by small sliding velocity, which promotes wear and, thus, the removal of the metal surface layer where fatigue cracks could appear.

An effective way of raising the contact endurance is to decrease the reduced curvature of a pair of active flanks at the common contact point. However, if the common point coincides with the pitch point, the reduced curvature of tooth profiles depends only on the pressure angle at the pitch point in any system of engagement under the given radial dimension of the pair. Regardless of the profile shape, the decrease of the reduced curvature by increasing the pressure angle can't be considerable because of the appearance of tooth thinning. Also, it should be taken into account that the rise of the pressure angle positively influences the contact endurance of the near-pitch-point area only in straight-tooth gearing. Helical gearing does not have such an effect, as with decreasing the reduced curvature of profiles, caused by the increase of the pressure angle simultaneously, the total length of contact lines is correspondingly shortened.

Taking into account the stated above, M.L. Novikov came to the following important statements as the result of his investigations:

1) If the common point of tooth contact does not coincide with the engagement pitch point, then for the helical gearing the convex addendum profile of one gear and the concave dedendum profile of the mating gear can be designed with the same curvatures, for example, circumscribing these profiles by a common

circular arc with the center at the pitch point. In this case, the reduced curvature of profiles appears to be equal to zero, the contact turns out to be linear and the whole profile becomes the contact line.

2) The gearing manufactured, according to the item 1, requires the ideal, precise location of profiles, which is almost always disturbed by center distance errors and elastic deflections of shafts under load. In order to compensate these radial errors, it is necessary to make the concave dedendum profile of one gear with curvature slightly less than of the convex addendum profile of the mating gear, having located the common contact point within the tangency of profiles. This gives the background for developing the theoretically point out-pitch-point engagement with the small reduced curvature of tooth profiles.
3) In order to provide high efficiency of the gearing, the common contact point should be located as near as possible to the pitch point of engagement, which, however, is restricted by the requirement of reducing the tooth addendum height because of the risk of its thinning.
4) Synthesis of a cylindrical gearing should be carried out in the following way: The position of the common contact point *K* is specified in the face plane, which coincides with the plane of the drawing (Fig. 1).

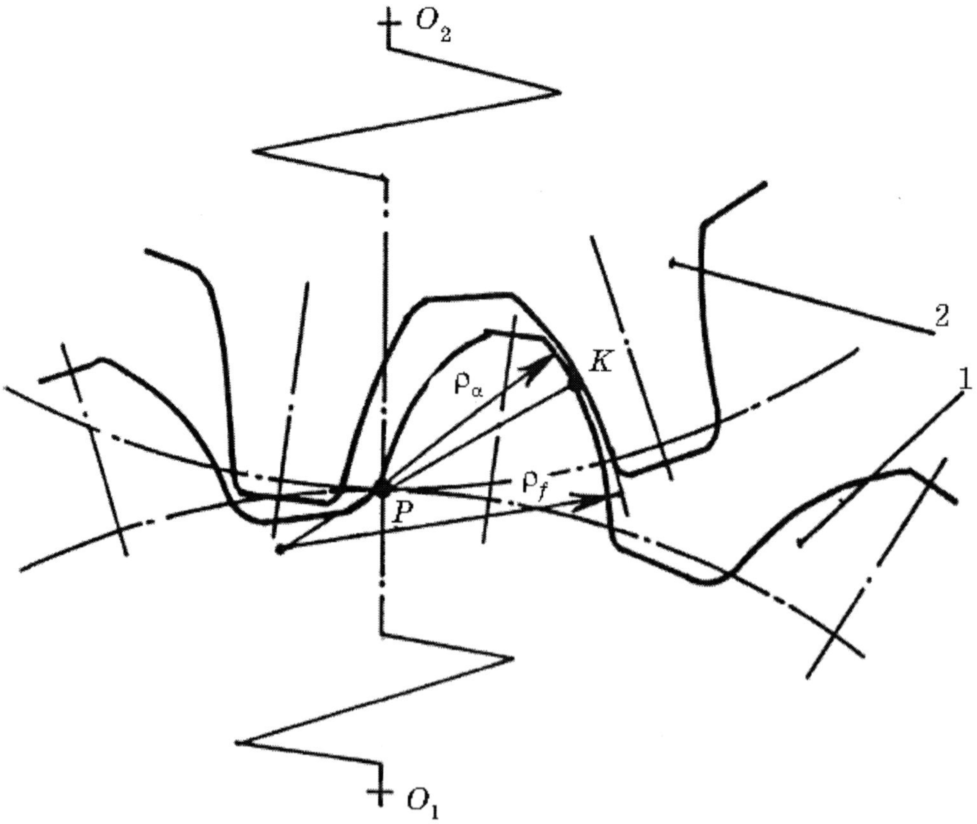

Figure 1. Scheme of Novikov gearing at the face plane.

Centers of profile curvature are chosen at the contact normal *PK* near the pitch point *P* and arcs are drawn through the point *K* with the radius ρ_a for the convex addendum of the gear *1* and ρ_f for the concave dedendum of the gear *2*. Unlike the conventional location of the line of action (at the face plane), the line of action is drawn through the point *K* in the three-dimensional space, for example, perpendicularly to the face plane. Then the law should be specified for the displacement of the common point *K* along the line of action under the rotation of gears around axes O_1 and O_2, assuming that, along with the point *K* the addendum and dedendum profiles are also displaced in three-dimensional space, their geometrical locus generating the contacting tooth flanks. The point *K* describes active contact lines on these flanks, at the same time the contact normal *PK* passes through the pitch point line at any moment of engagement. As a result, the tooth flanks of a non-straight-tooth gearing are generated. During operation of the gearing, proposed by M.L. Novikov, the point contacting under load is transformed into the contacting along the elastic area, occupying a considerable part of the tooth height and length. While gears are rotating, the elastic contact area is displaced along the tooth. Unlike straight-tooth gearing, for example, an involute one, where the continuity of engagement is provided by the face overlap of teeth, the continuity of engagement in Novikov gearing is achieved by the axial overlap.

Novikov gearing, synthesized in the described way, possesses the following advantages:

1) A relatively large contact area provides the high contact endurance, which significantly exceeds the endurance of the analogous involute gearing.
2) A relatively small distance from the line of action to the pitch point line enables to increase the efficiency of the gearing.
3) The local type of the contact improves dynamic characteristics and considerably reduces the gearing sensitivity to those manufacturing and operational errors, which cause load concentration along the tooth length in gearing, having a theoretically linear contact.
4) Displacement of the common contact point along the tooth, small longitudinal reduced curvature of tooth flanks and large rolling velocity of flanks in the direction of tooth lines increase the thickness of a hydrodynamic lubrication layer.
5) Inalterability of sliding velocities at any phase of engagement (for example, in cylindrical gearing with line of action parallel to the pitch point line) enables a good run-in of contacting flanks, reducing the non-uniform load and stress distribution between tooth pairs and further increasing the contact endurance.

Teeth of Novikov gears, like any other ones, can be cut by copying and generation methods. The monograph material concerns the gearing, having teeth manufactured by a more widely-spread practice generation method, when the tool is made according to the normal basic rack profile, corresponding to the normal tooth profile of the basic rack.

Among the whole scientific heritage of M.L. Novikov, the general equations are of special interest, which establish the relation between the gearing kinematics and tooth geometry at the vicinity of the common contact point and applied to gearing with any relative location of gear axes, any direction of line of action and any character of surface contacting. The well-known relationship of Euler-Savari, relating the curvatures of tooth profiles and gear centrodes in the plane engagement, is a particular case following from these equations. Convenient formulas for determination of the curvature of tooth flanks and orientation of the contact area are obtained on the base of Novikov equations [193].

The development of synthesis methods of out-pitch-point theoretically point engagement and grounding of the high contact strength of gearing, designed on the basis of such engagement is an outstanding achievement of M.L. Novikov, which received a high appraisal of native and foreign scientists. On the one hand, his research gave a powerful stimulus to the further development of general theory of the point engagement synthesis [64, 139, 193, 203], and, on the other hand, to the wide application of such engagement in industry [112]. Along with the authors of this monograph, a substantial contribution to the improvement of Novikov gearing, methods of its analysis and development of its production ws made by such Russian scientists as R.V. Fedyakin, V.A. Chesnokov, E.G. Roslivker, A.V. Pavlenko, V.N. Kudryavtsev, K.I. Zablonsky, Yu.F. Kouba, A.S. Yakovlev and others.

Chapter 1

VARIETIES OF BASIC RACK PROFILES FOR NOVIKOV GEARING AND THEIR DEVELOPMENT

The working capacity of Novikov gearing is mainly determined by the basic rack profile of gear teeth..

The term "basic rack profile of cylindrical gear teeth" denotes the tooth profile of a nominal basic gear rack in the section by a plane, perpendicular to its pitch plane. A normal basic rack profile is considered, usually, as a corresponding one to a profile in the section of a rack, normal to the tooth line.

Basic generating rack with teeththat are shaped in the form of slots of the basic rack, completely determines the shape of flank surfaces and tooth dimensions of gears. They are obtained as a result of generation (enveloping) in relative motion of the rack and the machined gear.

At the initial stage of NG development, gearing with one line of action (OLA) [158] was proposed, and then a standard basic rack profile according to MN 4229-63 [172] has been created as applied to teeth having surface hardness of tooth flanks under *HB 320*.

The basic rack profile, according to MN 4229-63, has small value $\Delta\rho$, that is the difference between radii ρ_f of the concave and ρ_a of the convex profiles of contacting surfaces, which predetermines the increased sensitivity of gearing with this profile to radial errors of gears during their manufacture and assembly. The basic rack profile OLZ-63 [196] with the increased value of $\Delta\rho$ and increased tooth height, which appeared shortly after, showed satisfactory results of fracture tests of very hard case-hardened teeth of a pair with a pinion having low tooth number and big gear ratio.

The important stage of NG development was the development of the gearing with two lines of action (TLA) [224].

Tests of TLA gearing with the basic rack profile Ural-2N and gear teeth hardness under *HB 320* [3] showed their undeniable advantages according to contact and fracture strength compared with OLA gearing.

Relying on the basic rack profile, Ural-2N, a standard GOST 15023-76, was developed for NG TLA regarding gears with tooth surface hardness under *HB 320,* module up to *16mm* and tangential speed up to *20 m/s* [45] using a standard series of modules especially developed for NG [37].

A great manufacturing advantage of TLA gearing compared with OLA gearing was the possibility to cut pinion and gear teeth by the same gear cutting tool.

Wide development of TLA gearing decreased considerably the application area of OLA gearing. The latter can be preferred only in some cases, when the number of pinion teeth is less than 8-10 and the gear ratio is big enough [196].

Attempts to achieve more uniform load distribution along the tooth led to proposal of NG with several (more than 2) lines of action.

For example, a gearing is assembled of gears with either two (convex and concave [262]), or one [72] of tooth profiles in normal section generated by certain number of arcs, obtained by enveloping, corresponding circular segments of the basic rack profile, connected by conversion curves.

However, it should be noted, that increasing active profile number of segments and lines of action for more than 2 is of low effectiveness for several reasons.

Firstly, the total height of an active profile, determining its workability according to the contact strength after run-in, is not more (and sometimes even less) than for TLA gearing, that is why the load-bearing capacity of such gearing is not increased.

Secondly, the presence of several segments of small height and, as a rule, with various pressure angles, inevitably leads to a rather high sensitivity of a gearing to manufacture errors, coming of contact areas outside the limits of active areas and to abrupt non-uniform distribution of the transmitted load between areas.

Thirdly, the described gearing requires very high manufacturing accuracy of the gearing itself and of a rather complicated tooth-cutting tool in order to harmonize the mutual arrangement of areas.

Evidently, for the noted reasons, the gearing with several lines of action haven't become widespread.

Considering the advantage and abundance of NG TLA, the attention is paid exactly to this gearing in this monograph.

During many years, a great number of basic rack profiles different in the shape and parameters has been proposed for TLA gearing, which can be divided into two main groups.

The first group comprises basic rack profiles with active arc segments of tooth addendum and dedendum connected by a rectilinear conversion segment. The well-known representatives of the first group are the basic rack profiles according to the standard GOST 15023-76 [45] and the profile YuTZ-65 [222], which are intended for gears with tooth surface hardness $H_{HB} \leq HB\,320$. The expanded active segments of teeth provide higher contact strength for these basic rack profiles, compared with involute analogues. It is the crucial factor for the given tooth hardness. Advantages of the basic rack profile YuTZ-65 are as follows: favorable vibroacoustic characteristic [26, 243] which is provided to a great extent by a uniform displacement of radii centers of the convex and concave profiles with respect to the pitch line and, consequently, by a theoretical conservation of both lines of action in operation under radial errors [101].

The principal drawback of the first group basic rack profiles is the presence of a rectilinear, involute-generating conversion segment within the near-pitch-point area, which participates in operation under definite conditions and becomes, as a rule, a source of the incipient pitting of surfaces [3]. Strictly, the presence of a conversion segment,which can be involved into operation within the pitch point zone, conflicts with the main idea of NG, which has been developed as out-pitch-point [158].

In view of stated above, and also because of the low fracture strength and increased sensitivity to errors, the application of NG with the noted basic rack profiles of the first group is practically eliminated for gears with hard tooth flanks.

As for basic rack profiles of the second group, the near-pitch-point area is out of operation due to the conversion concave segment, generated as a result of some shift of dedendum and addendum active profiles with respect to the axis of tooth symmetry.

As for basic rack profiles of the assemblage “Don” [196], such shift is comparatively small [84], and the concave conversion segment (which is the geometrical stress concentrator) is not hazardous here, that is proved by the performed tests.

Attempts to equalize bending stresses at the tooth root and in its middle part in order to increase the fracture strength, led to decisions [69, 182] to apply a considerable shift of addendum and dedendum segments, which is implemented, in particular, for basic rack profiles DLZ-0.7-0.15 and DLZ-1.0-0.15 [172]. Such a shift required the abrupt reduction of the tooth height down to $(0.6\ ...0.75)\ m$ and a reduction of the profile radius of the convex active segment, which negatively affected the contact (including depth) strength of the gearing and leads to the increase of its sensitivity to errors.

The evident increase of fracture strength by means of similar basic rack profiles is possible only by a high (not less than 5-6) degree of accuracy of gear manufacturing and assembly, that was proved by fatigue tests of pairs with carburized ground teeth with the basic rack profile DLZ-0.7-0.15 that showed the fracture strength *1.6* times greater than for involute analogues [172].

As for gearing with hard teeth of the middle degree of accuracy, the fracture strength of basic rack profiles with the decreased height and increased tooth root can not be achieved due to the enumerated drawbacks. The results of corresponding tests are described in details in Chapter 8.

Requirements to gearing of general engineering application with comparatively low (8-11) degrees of accuracy and with various tooth hardness can be met by the basic rack profile RGU-5 specially designed to rise the load-bearing capacity of such gearing [22].

The basic rack profile RGU-5 is notable for the combination of the shift x_a (performed within the given range) of the center of the tooth addendum curvature below the pitch line with a considerable inclination angle of the straight line with respect to the tooth axis of symmetry, connecting theoretical contact points of the addendum and dedendum, and also for a definite relation between the pressure angle at the point of contact and angles at points of conjugation of the conversion area and segments of the addendum and dedendum [59].

These features allow the obtaining of the tooth shape, close to a bar of uniform resistance, providing high fracture strength due to the enhanced tooth root in combination with favorable configuration and small curvature of near-pitch-point area. Sufficient tooth height and presence of the shift x_a provide high surface and depth contact strength under low sensitivity to manufacturing errors. Advantages of the basic rack profile RGU-5 are convincingly proved by long-term fatigue tests of nitro carburized ($H_{HRC} \geq HRC\ 56$) and thermally improved ($H_{HB} \leq HB\ 350$) gearing produced at Izhevsk PO “Reduktor”, and by the experience of its manufacture and operation (it is described in Chapters 8 and 9).

The development of basic rack profiles for NG TLA is visually illustrated in Fig.2.

Thorough analysis of theoretical and experimental investigations results showed that, in some cases, the fracture strength of gears with the basic rack profile RGU-5 is limited by a near-pitch-point geometrical concentrator. Thus, the basic rack profile RGU-5 was altered a little and it was the basis of development of the basic rack profile KS for gears with tooth hardness $H_{HRC} \geq HRC\ 35$and tangential velocity of gears up to *20 m/s*, developed by the Coordinating Board at USSR Minstankoprom on the problem of implementation of high-hardened NG with an active participation of authors [71, 264]. The basic rack profile KS and hobs for gear cutting with this basic rack profile were first standardized by industrial guidelines RD2 N24-11-88 [177] and RD2 I41-16-88, and on their basis, the Interstate standard GOST 30224-96 for the basic rack profilewas recently developed [39].

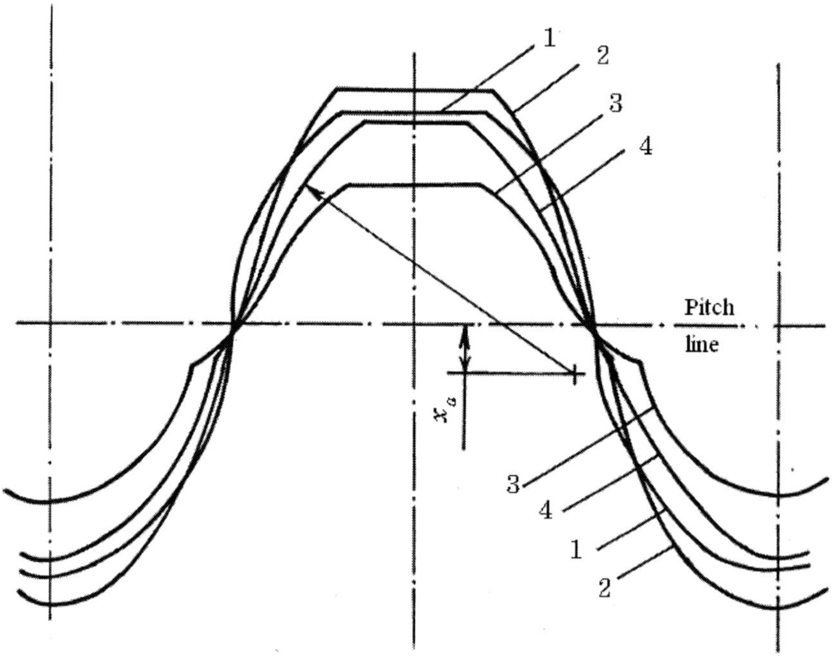

Figure 2. Development of basic rack profiles for Novikov gearing: 1 – according to the standard GOST 15023-76; 2 – Don-63; 3 – DLZ-0.7-0.15; 4 – RGU-5.

The feature of the basic rack profile according to the standard GOST 30224-96 is the presence of a small convex segment with radius $\rho_a^{'} < \rho_a$, conjugated with the active addendum and the conversion concave segment. This enables the rise of the fracture strength of the addendum by *(10...15) %* due to the increase of its thickness, keeping the fracture strength of the dedendum at practically the same level and keeping the required clearance within the near-pitch-point area for its secure elimination from operation.

For better visualization and convenience of performing calculations, schemes (Fig. 3-5) and tables of parameters (Tables 1-4) of the most widespread basic rack profiles are given below.

All the basic rack profiles considered above consist of a number of circular arcs, sometimes combined with straight lines, that facilitates the production of tooth-cutting tools. At the same time, a number of basic rack profiles generated by more complex curves is known; they are still not widely spread, but some of them may become useful as far as manufacturing techniques are advanced.

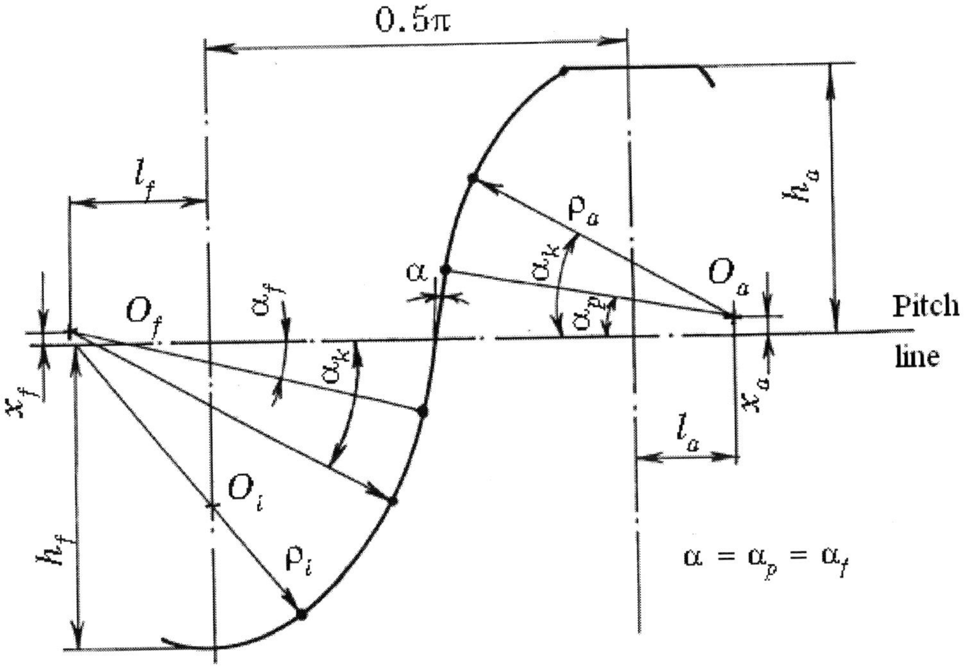

Figure 3. Scheme of TLA basic rack profile of the first group (according to the standard GOST 15023-76, YuTZ-65).

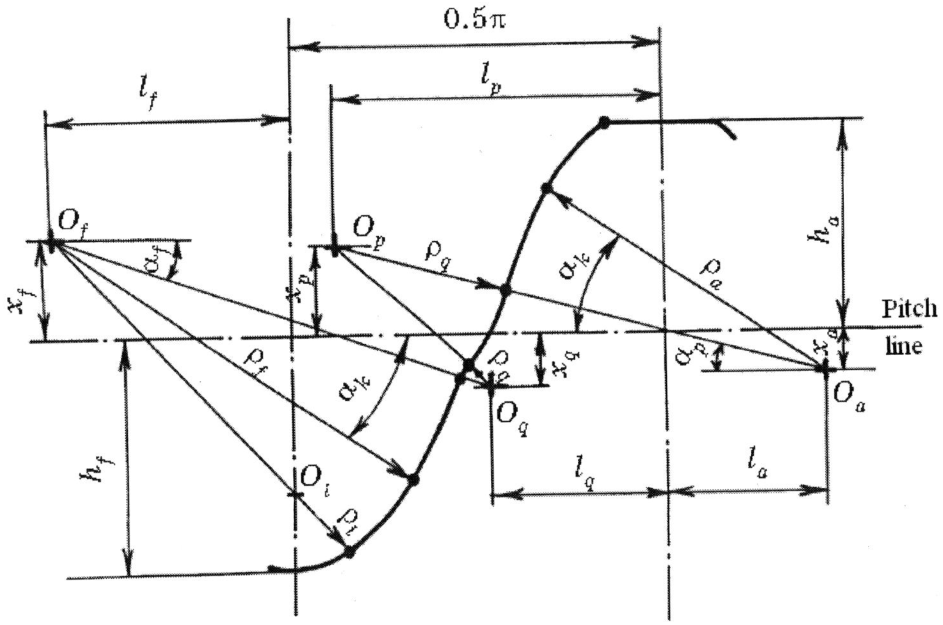

Figure 4. Scheme of TLA basic rack profile of the second group (Don-63, DLZ-0.7-0.15, DLZ-1.0-0.15, RGU-5).

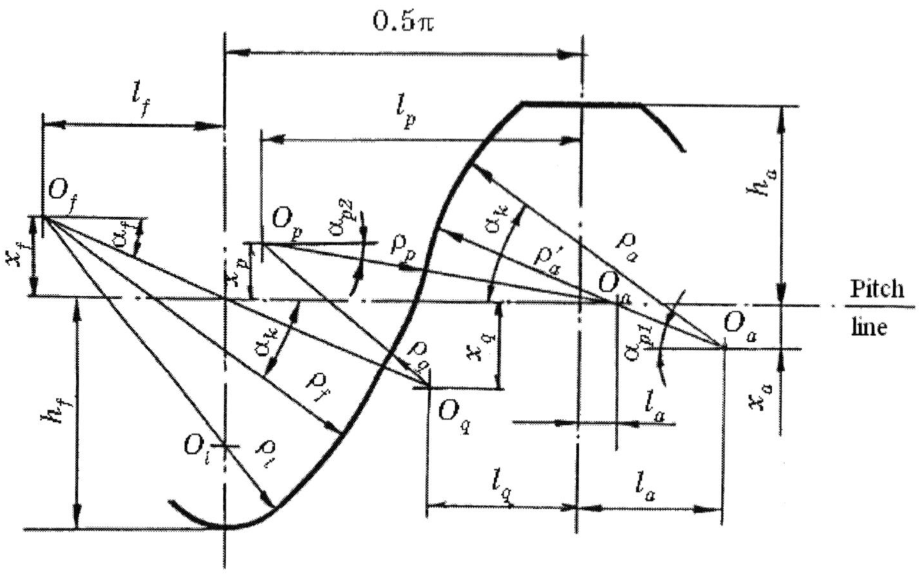

Figure 5. Scheme of KS basic rack profile (according to the standard GOST 30224-96).

Table 1. Main parameters of basic rack profiles of the first group

Parameter	According to the standard GOST 15023-76				YuTZ-65
	Module, mm				
	Up to 3.15	Over 3.15 to 6.3	Over 6.3 to 10	Over 10 to 16	
ρ_a^*	1.147	1.15			1.8
ρ_f^*	1.307	1.29	1.27	1.25	2.1
ρ_i^*	0.52246	0.52155	0.50677	0.49785	0.3115
h_a^*	0.9				1.26339
h_f^*	1.05				1.43661
x_a^*	0				-0.06339
l_a^*	0.3927				1.07981
x_f^*	0.07264	0.06356	0.05448	0.0454	0.06339
l_f^*	0.50526	0.48994	0.47462	0.4568	1.33652
$\alpha_k,^\circ$	27				25 *)
$\alpha,^\circ$	8.175	7.66306	8.38389	8.57694	12
j_k^*	0.06	0.055	0.05		0.03

*) Accepted conditionally.

Table 2. Main parameters of basic rack profiles Don-63, DLZ-0.7-0.15, DLZ-1.0-0.15.

Parameter	Don-63	DLZ-0.7-0.15	DLZ-1.0-0.15
ρ_a^*	1.5	0.7	1
ρ_f^*	1.85	0.85	1.15
ρ_p^*	0.5	0.6	0.55
ρ_q^*	0.08	0	
ρ_i^*	0.45	0.4	
h_a^*	1	0.6	0.75
h_f^*	1.19462	0.72772	0.9017
x_a^*	0		
l_a^*	0.8029	0.0746	0.3796
x_f^*	0.14792	0.08571	0.075
l_f^*	1.0786	0.1777	0.4795
x_p^*	0.27835	0.18092	0.31
l_p^*	1.17764	1.21275	1.13908
x_q^*	0.09634	-	
l_q^*	0.73492	-	
α_k, °	25	34.85	30
α_p, °	8	16.602	11.537
α_f, °	7.271	19.641	13.835
j_k^*	0.08302	0.04	0.06

Table 3. Main parameters of the basic rack profile RGU-5.

Parameter	Module, mm						
	2.5	3.15	4.0	5.0	6.3	8.0	10.0
ρ_a^*	1.41						
ρ_f^*	1.76	1.75		1.71	1.68	1.64	1.62
ρ_p^*	0.8346	0.71096	0.69962	0.83365	0.71095	0.71843	0.82874
ρ_q^*	0.2	0.15873	0.125	0.1	0.07937	0.0625	0.05

Table 3 (Continued).

ρ_i^*	0.32869	0.32506	0.32511	0.32809	0.32413	0.32352	0.3341
h_a^*	0.865						
h_f^*	1.00297	1.00377	1.00375	1.00101	1.00096	0.99922	0.99493
x_a^*	0.2						
l_a^*	0.71049						
x_f^*	0.39572	0.39013		0.36776	0.35098	0.32861	0.31743
l_f^*	0.95066	0.94236		0.9092	0.8843	0.85117	0.83459
x_p^*	0.43124	0.3668	0.36377	0.43097	0.3668		0.42959
l_p^*	1.44352	1.33333	1.32241	1.44261	1.33331	1.34053	1.4379
x_q^*	0.26123	0.24177	0.22032	0.21155	0.20242	0.19749	0.18938
l_q^*	0.67483	0.71206	0.74031	0.76521	0.78577	0.80279	0.81415
α_k, °	34						
α_p, °	16.333	15.5		16.333	15.5		16.333
α_f, °	19.583	19.333	19	18.667	18.333	18	17.667
j_k^*	0.1						

Table 4. Parameters of the basic rack profile S (according to the standard GOST 30224-96).

Parameter	Module over 2.5 to 4.5 mm	Module over 4.5 to 9.0 mm	Module over 9.0 to 16 mm
ρ_a^*	1.38		
ρ_f^*	1.76	1.66	1.56
$\rho_a'^*$	0.84611		
ρ_p^*	0.85		
ρ_q^*	0.05		
ρ_i^*	0.36558	0.36482	0.36386
h_a^*	0.875		
h_f^*	1.01163	1.00728	1.00238

Table 4 (Continued).

x_a^*	0.2		
l_a^*	0.64		
$l_a'^*$	0.14498		
x_f^*	0.41796	0.3606	0.30324
l_f^*	0.90128	0.81936	0.73745
x_p^*	0.32363		
l_p^*	1.51996		
x_q^*	0.14586	0.16117	0.17553
l_q^*	0.75213	0.7617	0.77108
$\alpha_k,°$	35		
$\alpha_{p1},° / \alpha_{p2},°$	22 / 11		
$\alpha_f,°$	18.15	17.766	17.3
j_k^*	0.1		

Thus, as for a gearing intended for big torques [70], reduction of sensitivity to errors is achieved due to the application of such tooth shape, for which active basic rack profiles are made with variable curvature, which is increased while moving away from the pitch surface.

Applying helixes, the curvature of which decreases while moving away from points of nominal contact, the sensitivity of the gearing becomes successfully low and the concentration of bending stresses in the most intense section of the dedendum also decreases [73].

NG was tested with the basic rack profile NIT [151, 252] based on the combination of the involute and circumference. Its dedendum contains the involute with the radius of curvature in the nominal contact point equal to the radius of addendum circle, that results in the decrease of sensitivity of the gearing compared with the one assembled of gears on the basis of a congruent pair of basic rack profiles.

Without discussing the great variety of published basic rack profiles, in conclusion, let's mention the basic rack profile Sym MarC of the company Hitachi [253], which is assigned in the face section for a standard tooth helix angle *15°*. Profiles of addendum and dedendum in this section are shaped by circular arcs of the same radius and centers of both arcs located at the initial line. In order to reduce the influence of center distance error on the load-bearing capacity of the gearing, shifts of basic rack profiles are applied to manufacture gears of the pair. It is shown in [263], that contact strength of gears with arc profiles three times exceeds the strength of involute gears, owing to this fact the proposed gears made of air-hardened steel successfully substitute involute gears with surface hardening of teeth.

Chapter 2

Geometry of Cylindrical Novikov Gearing

Problems of Novikov gearing geometry are described in literature in details.

There is a standard [44] for calculation of NG geometry; it concerns only gearing with the basic rack profile according to the standard GOST 15023-76, where teeth are cut without a shift of the basic rack profile.

Existing information is systemized in this Chapter, and also some new data is given on general-type gearing which gear teeth are cut with the shift of the basic rack profile [103].

2.1. Scheme of Gearing Conjugation

The scheme of gears conjugation in a gearing, described below, supposes a general case – the presence of a non-zero total shift of basic rack profiles when cutting the teeth of a pair of gears.

The necessity to apply gears, manufactured with the shift of the basic rack (they are called in literature "corrected" [249] or "non-zero" [18]), appeared first in the general gearbox industry, where a wide unification of gear pairs mounted at different stages of gearboxes is required [159, 223].

Schematically, the engagement of a pair of Novikov gears in the face plane, which are cut with the shift of the basic rack profile, is shown in Fig. 6 [249], where the common normal *nn* comprises: P – the pitch point of gears *1* and *2*; $P_1,$ P_2 - machine-tool pitch points for gears *1* and *2*, correspondingly; K – the theoretical contact point of the convex and concave tooth profiles of gears *1* and *2*, correspondingly. Normal *nn* is inclined to pitch planes (their traces are plotted on the plane of drawing by lines *1* and *2*) by the angle α_{kt}, which is related to the angle α_k of a normal basic rack profile by the known dependence:

$$tan\alpha_{kt} = tan\alpha_k / cos\beta, \tag{2.1}$$

where β is the helix angle of the rack (or tooth of a gear at the pitch cylinder).

The angle δ, obtained for the total shift $x^*_\Sigma \neq 0$ is equal to

$$\delta = arctan(O_1O_1^{'} / O_1^{'}O_2) = arctan(O_2O_2^{'} / O_1O_2^{'}) = arctan\{x_{\Sigma}^{*} / [(a^{*} + x_{\Sigma}^{*})tan\alpha_{kt}]\},$$

where, in turn, the pitch center distance a^{*} is the sum of a pair of gears pitch radii:

$$a^{*} = r_1^{*} + r_2^{*} = (z_1 + z_2)/(2cos\beta). \tag{2.2}$$

Using the formula (2.1), we obtain

$$\delta = arctan\{x_{\Sigma}^{*} cos\beta / [(a^{*} + x_{\Sigma}^{*})tan\alpha_k]\}. \tag{2.3}$$

Now the center distance a_w^{*} of a pair of gears is determined easily:

$$a_w^{*} = O_1O_2 = [(a^{*} + x_{\Sigma}^{*})^2 + (x_{\Sigma}^{*} cos\beta / tan\alpha_k)^2]^{0.5}, \tag{2.4}$$

also, the face pressure angle α_{ktw}(in a theoretical point *K* of profiles contact)

$$\alpha_{ktw} = \alpha_{kt} + \delta = arctan(tan\alpha_k / cos\beta) + arctan\{x_{\Sigma}^{*} cos\beta / [(a^{*} + x_{\Sigma}^{*})tan\alpha_k]\}. \tag{2.5}$$

If necessary, the normal pressure angle can be determined (in point *K*):

$$\alpha_{kw} = arctan(tan\alpha_{ktw} cos\beta_w), \tag{2.6}$$

where the helix angle β_w of a gear at the reference cylinder is

$$\beta_w = arctan(a_w^{*} tan\beta / a^{*}). \tag{2.7}$$

Calculations show, that the angle δ is usually rather small, and it often enables to apply in practice, the approximating relationship for center distance calculation:

$$a_w = m(a^{*} + x_{\Sigma}^{*}). \tag{2.8}$$

In those cases, when the center distance a_w is given for design, the coefficient x_{Σ}^{*} of the total shift is determined according to the formula:

$$x_{\Sigma}^{*} = x_1^{*} + x_2^{*} = \left[\sqrt{(a_w^{*2} - a^{*2})cot^2\alpha_k cos^2\beta + a_w^{*2}} - a^{*}\right] / (1 + cot^2\alpha_k cos^2\beta). \tag{2.9}$$

The question of reasonable distribution of x_{Σ}^{*} between the pinion (x_1^{*}) and gear (x_2^{*}) in order to increase the contact strength of a gearing is described in [213].

2.2. Main Geometrical Dimensions of Gears

Pitch $d_{1,2}$ and reference $d_{w1,2}$ diameters of gears are determined according to these known formulas:

$$d_1 = mz_1 / \cos\beta; \qquad d_2 = mz_2 / \cos\beta; \tag{2.10}$$

$$d_{w1} = 2a_w /(u+1); \qquad d_{w2} = 2a_w u /(u+1), \tag{2.11}$$

where $u = z_2 / z_1$ is the gear ratio of the pair.

When determining tooth tip and root diameters, no compensating shift is introduced, as it is provided for the involute engagement [42], that is why tooth tip (d_a) and root (d_f) diameters are:

$$d_{a1,2} = d_{1,2} + 2m(h_a^* + x_{1,2}^*), \tag{2.12}$$

$$d_{f1,2} = d_{1,2} + 2m(x_{1,2}^* - h_f^*). \tag{2.13}$$

The nominal height of the gear tooth is

$$h = m(h_a^* + h_f^*), \tag{2.14}$$

And the nominal radial backlash in the pair is

$$c = m(h_f^* - h_a^*). \tag{2.15}$$

2.3. Multiple-Point Contact, Total and Phase Factors of Tooth Axial Overlap

According to many references (for instance, [172]), it is known that the contacting of teeth of NG TLA takes place in several points, and when the engagement phase (angle of the pair rotation) is changed, the number of contact points also varies from a certain minimum to maximum. For strength analysis, the most valuable information is the most unfavorable case – a minimum theoretical number of contact points in the engagement of a pair, which depends both on the axial overlap ratio ε_β for each of two lines of action, and on the distance q between projections on the pitch point line of two theoretical contact points following one after another, located on different lines of action.

The axial overlap ratio is known to be expressed by the formula

$$\varepsilon_\beta = b_w / p_x, \tag{2.16}$$

where b_w is the operating width of toothing of a pair, equal to minimum from b_1 (pinion toothing width) and b_2 (gear toothing width), p_x is the axial pitch, equal to $\pi m / \sin\beta$.

As for the distance q, the relation for its determination is described in [129, 239]:

$$q^* = max\{q_{21}^*, q_{22}^*\}, \tag{2.17}$$

where $q_{22}^* = F^* \sin\beta + G^* / \sin\beta$, $q_{21}^* = p_x^* - q_{22}^*$.

Constants F^* and G^* are expressed by parameters of the basic rack profile: $F^* = 2\cos\alpha_k(\rho_a^* - x_a^* / \sin\alpha_k)$, $G^* = 0.5\pi - l_a^* - l_f^* + (x_a^* + x_f^*) / \tan\alpha_k$.

Dividing the considered parameters to the axial pitch p_x^*, we obtain the design axial range of contact points, or the phase axial overlap ratio:

$$\varepsilon_q = max\{C_q^*, (1 - C_q^*)\}, \tag{2.18}$$

where $C_q^* = A_q^* \sin^2\beta + B_q^*$, $A_q^* = F^* / \pi$, $B_q^* = G^* / \pi$.

Coefficients A_q^*, B_q^* depend on parameters of the basic rack profile and in order to simplify calculations, they can be taken from the Table 5.

Table 5. Coefficients A_q^*, B_q^*.

Coefficient	Basic rack profile				
	RGU-5	According to the standard GOST 30224-96	According to the standard GOST 15023-76*)	DLZ-0.7-0.15	DLZ-1.0-0.15
A_q^*	0.55541	0.53782	0.65232	0.36571	0.55133
B_q^*	0.25237	0.29032	0.25875	0.45887	0.26789

*) for the module over 3.15 to 6.3 mm.

On the basis of the performed investigations, the following Table 6 can be proposed to determine the minimum number n_{min} of theoretical contact points.

It is interesting to note, that the case $\varepsilon_q = 0.5$, i.e., when the neighboring points of different lines of action are retarded from each other exactly by the half of the axial pitch, is a special one, since for some values ε_β gives n_{min} more than for $\varepsilon_q \neq 0.5$ by the value 1 (Table 6).

Table 6. Minimum number n_{min} of theoretical contact points which are in the engagement simultaneously.

Range ε_β	n_{min}	
	when $\varepsilon_q \neq 0.5$	when $\varepsilon_q = 0.5$
$\varepsilon_\beta < \varepsilon_q$	0	0
$\varepsilon_q \leq \varepsilon_\beta < 1$	1	1
$1 \leq \varepsilon_\beta < 2 - \varepsilon_q$	2	2
$2 - \varepsilon_q \leq \varepsilon_\beta < 1 + \varepsilon_q$	2	3
$1 + \varepsilon_q \leq \varepsilon_\beta < 2$	3	3
$2 \leq \varepsilon_\beta < 3 - \varepsilon_q$	4	4
$3 - \varepsilon_q \leq \varepsilon_\beta < 2 + \varepsilon_q$	4	5
$2 + \varepsilon_q \leq \varepsilon_\beta < 3$	5	5
$3 \leq \varepsilon_\beta < 4 - \varepsilon_q$	6	6

2.4. Reduced Main Contacting Tooth Flanks Curvature Radii

Reduced main contacting tooth flanks curvature radii of a pair of gears at a theoretic contact point, which are necessary to know for performing strength analysis, are usually calculated according to the following formulas [140, 176]: - in the direction of the tooth height (profile radius)

$$\rho_\alpha^* = \rho_a^* \rho_f^* / (\rho_f^* - \rho_a^*); \tag{2.19}$$

- in the direction of the tooth line (longitudinal radius)

$$\rho_\beta^* = z_2 / [2(u+1) \sin \alpha_k \sin^2 \beta \cos \beta]. \tag{2.20}$$

It should be noted that values of radii ρ_α^* and ρ_β^* calculated according to formulas (2.19) and (2.20) are approximate. Precise definition of these radii and calculation of the instant elliptical contact area angle of rotation with respect to the direction of the tooth line of a gear can be performed, if necessary, according to the engineering algorithm, composed on basis of works [125, 193] and given in Appendix 1.

Calculations showed that, for gears, based on basic rack profiles with $x_a^* = 0$ and with shifts x_1^*, x_2^* within the assigned restrictions (they are described in Chapter 3), the deviation of approximate values ρ_α^* and ρ_β^* from precisely calculated is not big and it is practically of no influence on results of strength analysis. For gears with basic rack profiles $x_a^* > 0$, this deviation is more noticeable, which is taken into account in strength analysis by means of correction factors (see Chapter 11).

2.5. Backlash in Gearing

It is known that in a zero gear (when $x_\Sigma^* = 0$), the backlash is equal to the assigned one in the basic rack profile [239], and it is considered to be practically the same as a non-zero gear [155]:

$$j_k = 2m\left[(\rho_f^* - \rho_a^*)\cos\alpha_k - l_f^* + l_a^*\right], \tag{2.21}$$

where the meaning of parameters included into (2.21) is clear from Fig. 3-5.

Turning to Fig. 6, we see that the rotation of the gear *1*, necessary for the left contact normal to pass to the right one (or vice versa) at the addendum, is determined by the angle

$$\kappa_1 = 4\tilde{S}_1^* / z_{v1} - 2\delta, \tag{2.22}$$

where $z_{v1,2} = z_{1,2} / \cos^3\beta$ is the reduced number of teeth of an equivalent straight-tooth gear, $\tilde{S}_1^* = l_a^* - (x_a^* - x_1^*)\cot\alpha_k$ is the arc of rotation at the pitch circle.

The corresponding rotation of the gear *2*

$$\kappa_2' = \kappa_1 z_{v1} / z_{v2} = 4\tilde{S}_1^* / z_{v2} - 2\delta \cdot z_{v1} / z_{v2}. \tag{2.23}$$

The angle of gear *2* rotation, necessary for the left contact normal of the dedendum to pass to the right one, is equal to

$$\kappa_2 = 4\tilde{S}_2^* / z_{v2} - 2\delta, \tag{2.24}$$

where the arc $\tilde{S}_2^* = l_f^* - \left(x_f^* + x_2^*\right)\cot\alpha_k + 0.5\pi$.

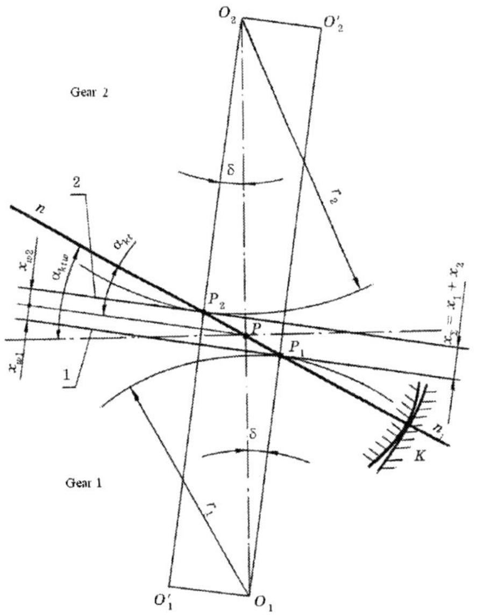

Figure 6. Scheme of engagement at the face plane of a pair of Novikov gears, cut with the shift of basic racks.

Taking into account that the addendum of the gear *1* is between dedendums of two neighboring teeth of the gear *2*. which are at the distance of the angular pitch $2\pi / z_{v2}$ from each other, we obtain the angle of the gear rotation, necessary to the backlash adjustment ("dead stroke"):

$$\lambda_2 = \left| \kappa_2' - (\kappa_2 - 2\pi / z_{v2}) \right| = 2\left(j_k^* + \Delta^*\right) / z_{v2}. \tag{2.25}$$

(Due to a small value of the angle δ, it is taken that $tan\ \delta = \delta$ for transformations).

In the formula (2.25), the value Δ^*, appearing for $x_\Sigma^* \neq 0$ and equal to

$$\Delta^* = 4(x_\Sigma^*)^2 \cot\alpha_k / \left(z_{v1} + z_{v2} + 2x_\Sigma^*\right), \tag{2.26}$$

is added to the backlash j_k^* (2.21),

Considering the survey from the gear (addendum) to the pinion (dedendum), we obtain the similar result, which means that for $x_\Sigma^* \neq 0$, both lines of action are kept theoretically.

It follows from (2.26), that independently of the sign of x_Σ^*, the backlash

$$j_w = m\left(j_k^* + \Delta^*\right) \tag{2.27}$$

in a non-zero gearing is always more, than j_k, and due to $\left|x_\Sigma^*\right| << \left(z_{v1} + z_{v2}\right)$ the function $\Delta^* = f(x_\Sigma^*)$ is close to a quadratic parabola with minimum $\Delta^* = 0$ for $x_\Sigma^* = 0$.

Chapter 3

QUALITY PARAMETERS FOR CYLINDRICAL NOVIKOV GEARING

Tooth geometry of a gear, cut by the generation, differs from the basic rack profile geometry (tooth of a nominal basic rack), this difference being greater for smaller numbers of gear teeth and for greater shift coefficient of the basic rack profile when cutting.

Similarly to involute gearing, there are factors of geometrical nature for NG, imposing limitations on parameters of the basic rack profile and gear [19, 104, 105, 146]. To determine these factors is to make essentially the quality control of engagement, which should be carried out when designing a gear. Letus assume the following characteristics as the main parameters of the engagement quality: addendum undercut, thinning of the tooth tip, location of the pitch point line with respect to active areas of the tooth and the degree of decreasing the active height of the tooth related to its cutting by the generation [117]. The pointed parameters are described below.

3.1. ADDENDUM UNDERCUT

From a mathematical point of view, the tooth undercut is the appearance of singular points on tooth flanks, their locus generating the so-called edge of regression, where the surface regularity (smoothness) is broken. In singular points, the position of the plane, tangent to the surface, is not determined since the so-called self-intersection of surfaces takes place [146].

The undercut has a negative influence both on fracture strength of teeth, creating concentrators of increased stresses, and on contact strength, decreasing the extent of the active area with respect to height and abruptly increasing the curvature of a surface near singular points. Therefore, the phenomenon of tooth flanks undercut should be eliminated, if possible.

It is known that in NG TLA, the dedendum of a gear doesn't undergo undercut (unlike involute teeth), since it has a concave profile, however, during the generation, the thickness of its "risky" section can be significantly decreased for small numbers of gear teeth [105, 115]. This factor is not considered here, because modern analysis methods of tooth fracture (Chapter 13) take it into account by themselves. Note also, that in gears, made on the basis of the first group basic rack profiles, when the shift of the basic rack profile is negative, the conversion involute segment is also undercut, however, the addendum undercut appears earlier and becomes a limiting factor.

Taking into account all the described above, letus consider the undercut of the convex addendum of the Novikov gear.

The phenomenon of addendum undercut was studied in [18, 146] and the condition of the undercut was obtained there in the form of the equation

$$sin^4 \vartheta_n + a_v^{'} sin^3 \vartheta_n + b_v^{'} sin \vartheta_n + c_v^{'} = 0, \quad (3.1)$$

where $a_v^{'} = z/(2\rho_a^* sin^2 \beta cos\beta) - (x_a^* - x^*)/\rho_a^*$, $b_v^{'} = (x_a^* - x^*)/(\rho_a^* tan^2 \beta)$, $c_v^{'} = -[(x_a^* - x^*)/(\rho_a^* tan\beta)]^2$.

Defining real roots of the equation (3.1), the angle of the undercut ϑ_n is obtained.

The solution of the biquadratic equation (3.1) requires the application of a computer. Our investigations led to the essential simplification of this task and to the possibility to apply the cubic equation of the following type without loss of accuracy

$$sin^3 \vartheta_n + a_v sin \vartheta_n + b_v = 0, \quad (3.2)$$

where $a_v = 2(x_a^* - x^*)/z_v$; $b_v = -2(x_a^* - x^*)^2/(z_v \rho_a^*)$.

The undercut of the active part of the addendum will not occur if the condition is fulfilled:

$$\vartheta_n \le \alpha_p. \quad (3.3)$$

For gearing with the basic rack profile, according to the standard GOST 30224-96, a certain undercut of an additional convex segment can be allowed in most cases, if the angle α_p is taken to be equal to (see Fig. 5 and Table 4)

$$\alpha_p = 0.56(\alpha_{p1} + \alpha_{p2}), \quad (3.4)$$

and in critical cases, it is recommended to perform the check of the absence of undercut of an additional convex addendum segment by replacing the value $\rho_a^{'*}$ instead of the radius ρ_a^* into (3.2) and (3.3) and accepting $x_a^* = 0$ and $\alpha_p = \alpha_{p2}$.

Expanding coefficients a_v, b_v and solving (3.2), with respect to the shift coefficient x^*, we obtain simple engineering formulas for limiting values: $x_{min}^* = x_a^* - 0.5\rho_a^* sin\alpha_p(L+1)$,

$$x_{max}^* = x_a^* + 0.5\rho_a^* sin\alpha_p(L-1), \quad (3.5)$$

where $L = \sqrt{1 + 2 sin\alpha_p z_v / \rho_a^*}$.

For the additional segment of the tooth when the basic rack profile is taken according to the standard GOST 30224-96, the application of the formula (3.5) is possible when ρ_a^* is substituted for $\rho_a'^*$, $x_a^* = 0$ and $\alpha_p = \alpha_{p2}$.

Now the condition of the undercut absence can be presented as

$$x^*_{min} \le x^* \le x^*_{max}. \tag{3.6}$$

Accepting, as it was mentioned above, a certain undercut of the additional segment for the basic rack profile according to the standard GOST 30224-96, we obtain the dependence for the absolute absence of undercut instead of (3.6), its breaking will lead to a considerable worsening of gearing operation

$$1.4x^*_{min} \le x^* \le 1.4x^*_{max}. \tag{3.7}$$

As an example, Fig. 7 presents the diagram for the conditional absence of undercut (3.6) (dotted lines), and the diagram for the absolute absence of undercut (3.7) (solid lines) for the basic rack profile according to the standard GOST 30224-96.

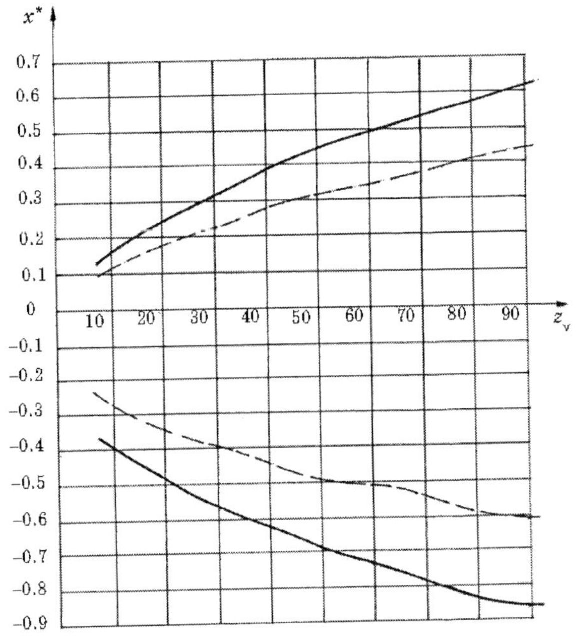

Figure 7. Diagram of dependence of coefficients x^* of the shift of the basic rack profile on the number z_v of teeth of the equivalent straight-tooth gear for the absence of undercut for the basic rack profile according to the standard GOST 30224-96: conditional absence of undercut (dotted lines); absolute absence of undercut (solid lines).

3.2 Thinning of Tooth Tip

It is known from literature, that the tooth with a very small thickness, S_a of the tip margin, is subjected to chippage of the tip, especially when the material hardness is high. Experience of tests and operation of Novikov gears [239] approximately showed that the allowable value of tooth thickness at the tip can be taken $[S_a^*] = 0.2$ for teeth with a homogeneous material structure and $[S_a^*] = 0.32$ for teeth with surface hardening up to high hardness.

The known engineering formulas [19, 239], proposed for definition of S_a, give satisfactory results within the limited range of geometrical parameters. That is why we developed a computational program for definition of S_a according to a precise technique, based on the calculation of the point of intersection coordinates of the gear addendum surface by a cylindrical outer surface.

In a planar case (which is rather correct), the following equation can be used:

$$\left(x_S^*\right)^2 + \left(y_S^*\right)^2 = r_{av}^2, \tag{3.8}$$

where x_S^*, y_S^* are the coordinates of the noted point of intersection, r_{av} is the tip radius of an equivalent spur gear, equal to $0.5\, z_v + h_a^* + x^*$.

Coordinates x_S^*, y_S^* are connected with parameters of the basic rack profile by the current angle ϑ of the rack tooth profile and the angle φ_S of run-in:

$$\begin{cases} x_S^* = A_S^* \cos\varphi_S + B_S^* \sin\varphi_S, \\ y_S^* = B_S^* \cos\varphi_S - A_S^* \sin\varphi_S, \end{cases}$$

$$A_S^* = \rho_a^* \sin\vartheta + \left(x^* - x_a^*\right) + 0.5 z_v,$$

$$B_S^* = \rho_a^* \cos\vartheta + \left(x^* - x_a^*\right)\cot\vartheta,$$

$$\varphi_S = 2\left[\left(x^* - x_a^*\right)\cot\vartheta + l_a^*\right] / z_v.$$

Then

$$S_a^* = 2\, y_S^*, \tag{3.9}$$

and the condition of the tooth thinning absence is

$$S_a^* \ge [S_a^*] \tag{3.10}$$

Numerical simulation with the help of the created program allowed to obtain the approximating dependence

$$x_{max}^* = a_S\left(z_v - b_S\right)^{\gamma_S}, \tag{3.11}$$

where coefficients a_S, b_S, γ_S depend on parameters of the basic rack profile and they are taken according to the Table 7.

Table 7. Values of coefficients a_S, b_S, γ_S.

Coefficient	Gearing with the basic rack profile RGU-5		Gearing with the basic rack profile according to the standard GOST 30224-96	
	Teeth of material with homogeneous structure	Teeth with surface hardening	Teeth of material with homogeneou s structure	Teeth with surface hardening
a_S	0.19	0.094	0.25	0.106
b_S	5	7	5	5
γ_S	0.59	0.65	0.54	0.67

Now the condition of the tooth thinning absence can be expressed as

$$x^* \le x^*_{max}. \tag{3.12}$$

As the results showed, tooth thinning will not occur for gears with the basic rack profile according to the standard GOST 15023-76, and as for basic rack profiles RGU-5 and those taken according to the standard GOST 30224-96, it can occur only for the very small tooth number z. In vast majority of cases, the addendum undercut (even the absolute one for the basic rack profile according to the standard GOST 30224-96) is of more severe limitation than thinning.

3.3. Location of Pitch Point Line

For the whole number of modern basic rack profiles (in particular, basic rack profiles of the second group – see Chapter 1), the unfavorable from contact point of view near-pitch-point zone is out of operation by means of a concave conversion segment *ab* (Fig. 8), that is especially important for teeth with high-hardened tooth flanks.

For gearing, cut with shifts x^*, such a situation is possible in the engagement, when the location of a pitch point line is within the active areas of addendum or dedendum surfaces, that is, in the zone of contact, which is inadmissible, since the duty of the segment *ab* for the pitch point elimination becomes useless.

According to Fig. 8, we have:

$$t_a^* = \rho_a^* \sin\alpha_p - x_a^*,$$
$$t_f^* = \rho_f^* \sin\alpha_f - x_f^*. \tag{3.13}$$

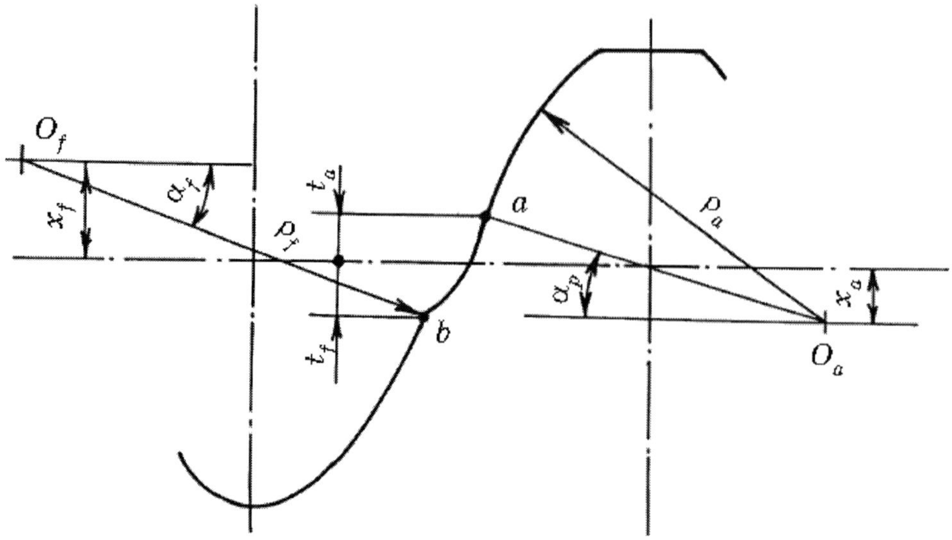

Figure 8. To define the allowable location of the pitch point line in the gearing.

In order to determine the value of the pitch point line shifting, let's turn to Fig. 6, where x_{w1} and x_{w2} are the distances from the pitch point line to pitch lines *1* and *2*, correspondingly.

Since $x^*_\Sigma = x^*_1 + x^*_2 = x^*_{w1} + x^*_{w2}$, and the pitch point P of the gearing is always located from pitch lines *1* and *2* at distances, proportional to teeth numbers z_1 and z_2, we easily obtain

$$x^*_{w1} = z_1 x^*_\Sigma / (z_1 + z_2), \qquad x^*_{w2} = z_2 x^*_\Sigma / (z_1 + z_2). \tag{3.14}$$

Then the required value of the pitch point line shifting is

$$x^*_w = x_1 - x^*_{w1} = x^*_{w2} - x_2 = x_1 - z_1 x^*_\Sigma / (z_1 + z_2). \tag{3.15}$$

Now the condition of the allowable pitch point line location in a gearing is expressed as

$$\left| x^*_w \right| \le \left[x^*_w \right], \tag{3.16}$$

where $[x^*_w] = min\{t^*_a, t^*_f\}$.

Note that a considerable shifting of the pitch point line, as it follows from (3.15), appears in a gearing for $x^*_1 = -x^*_2 \quad (x^*_\Sigma = 0)$.

3.4. Decrease of the Tooth Active Height, When Cut by Generation

It is known that in order to determine the load-bearing capacity of NG, according to contact endurance of tooth flanks with account of their run-in, it is necessary to know the design value of the addendum active segment [129, 239]. It has been determined according to the basic rack profile with introducing some empirical corrections for a gear tooth so far [129]. Such a situation can not be declared satisfactory [199], that is why the task was set and solved on definition of an actual extension of the tooth active area (of the gear tooth, cut by generation) with respect to height.

Letus designate the extension (chordal length of arc) of the tooth active area with respect to height as l, the length of an active area of the gear tooth as l_z, then the difference of l_z from l can be evaluated by the coefficient

$$K_l = l_z / l. \tag{3.17}$$

According to Fig. 9, the value l^* is calculated by the formula

$$l^* = 2\rho_a^* \sin[0.5(\alpha_a - \alpha_p)] \tag{3.18}$$,

α_a is the profile angle at the tooth tip, equal to

$$\alpha_a = \arcsin[(h_a^* + x_a^*) / \rho_a^*]. \tag{3.19}$$

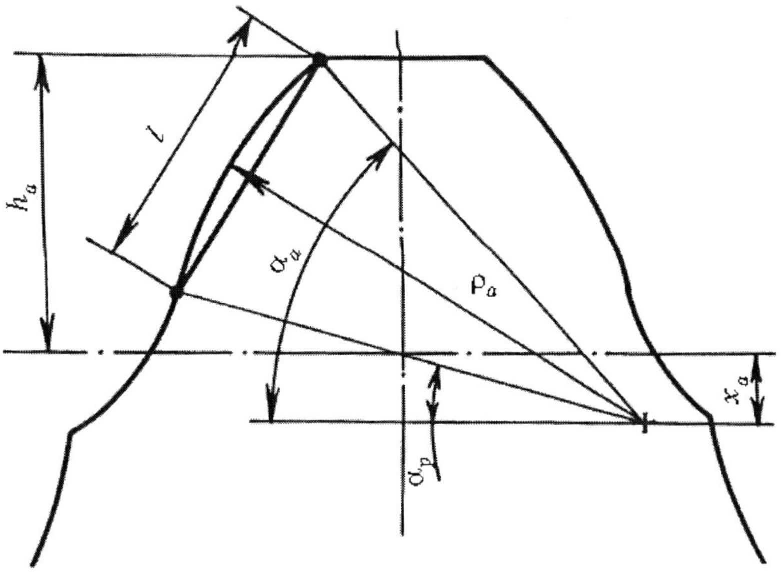

Figure 9. To define the chordal length of the active area of the basic rack addendum profile.

As calculations showed, if the basic rack profile is taken according to the standard GOST 30224-96, where a certain undercut of an additional convex segment is allowed, the value l^* can be taken equal to 0.776 regardless the module.

The tooth profile of the basic rack (also as the tooth profile of the gear) has a complicated shape, consisting of several segments.

Letus show the definition of the coefficient K_l by the example of cutting the teeth with the basic rack profile according to the standard GOST 30224-96.

During the generation (cutting), not only can the interference of the 1st type (undercut of convex segments) occur, but also intersections of various segments with each other and with the loss of active surfaces regularities. This phenomena can lead to reduction of active segments, which has to be taken into account. Thus, Fig. 10 shows the tooth profile in the coordinate plane *xOy* of the gear, cut by generation. Digits *1-6* designate segment numbers.

During the generation, the "surplus" area (shaded) was obtained, which is truncated. When segments *2* and *3* intersect each other, a certain active part of an additional segment *2* is lost (exaggeratedly shown).

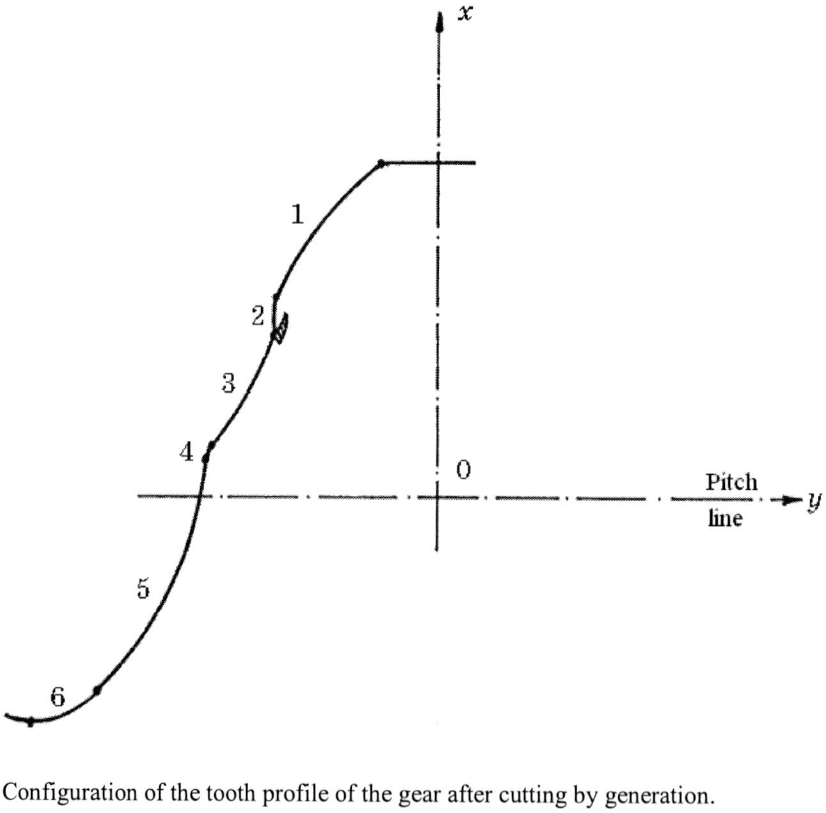

Figure 10. Configuration of the tooth profile of the gear after cutting by generation.

Considering the generation task in planar statement (that is, for an equivalent straight-tooth gear) without loss of accuracy, letus designate coordinates of points of the gear tooth profiles at the addendum and dedendum as $x_{t,r}, y_{t,r}$ correspondingly, they are shaped in the generation at the current angle of the basic rack enveloped profile, equal to a theoretical pressure angle of the gearing.

Letus introduce the double subscripting *ij*, where *i* is the number of the segment of the enveloped profile, $j=1...n$ is the number of the point at each segment when dividing it by *n* parts. Then, the following expressions of coefficients can be presented, meaning the relation of corresponding segments of gear tooth profiles to the value *l* of the basic rack: - for the active segment of addendum profile from the point x_t, y_t towards the increase of profile angles $K_A = l^{-1}\left[(x_{11}-x_t)^2+(y_{11}-y_t)^2\right]^{0.5}$; - for the active segment of addendum profile from the point x_t, y_t towards the reduction of profile angles $K_P = l^{-1}\left\{\left[(x_{1n}-x_t)^2+(y_{1n}-y_t)^2\right]^{0.5}+0.5\left[(x_{1n}-x_{2n})^2+(y_{1n}-y_{2n})^2\right]^{0.5}\right\}$;- for the active segment of dedendum profile from the point x_r, y_r towards the reduction of profile angles $K_F = l^{-1}\left[(x_{51}-x_r)^2+(y_{51}-y_r)^2\right]^{0.5}$; - for the active segment of dedendum profile from the point x_r, y_r towards the increase of profile angles $K_I = l^{-1}\left[(x_{5n}-x_r)^2+(y_{5n}-y_r)^2\right]^{0.5}$.

The additional component in the expression for the coefficient K_P considers the half of the additional segment *2* within the active part of the addendum.

In order to determine the pointed coefficients, a computational program "Obkatka" was specially developed. By means of this program, based on statements of the theory of gearing and differential geometry, the table for the choice of coefficients K_A, K_P, K_F, K_I within the range $z_v = 9...100$ was developed.

The following approximating polynomial for computation of the noted coefficients was obtained by numerical simulation

$$\sum_0^2 A_{0q}\left(x^*+1.5\right)^q + z_v^{-1}\sum_0^4 A_{1q}\left(x^*+1.5\right)^q + z_v^{-2}\sum_0^4 A_{2q}\left(x^*+1.5\right)^q + \\ + z_v^{-3}\sum_0^4 A_{3q}\left(x^*+1.5\right)^q, \tag{3.20}$$

it gives the error not more than *2%* compared with results of computations by the program "Obkatka" and it is intended for application in computer-aided algorithms for strength analysis.

Coefficients, included into the polynomial (3.20) are taken from special tables. As an example, Table 8 is given for gearing with the basic rack profile according to the standard GOST 30224-96.

Now we obtain: - for the part of engagement, approaching the pitch point $K_l^{(1)} = min\{K_{A1}, K_{I2}\} + min\{K_{P1}, K_{F2}\}$; - for the post pitch point part of engagement $K_l^{(2)} = min\{K_{A2}, K_{I1}\} + min\{K_{P2}, K_{F1}\}$.

The final coefficient is determined as

$$K_l = min\left\{K_l^{(1)}, K_l^{(2)}\right\}. \tag{3.21}$$

For the convenience of engineering calculations, tables of coefficients K_A, K_P, K_F, K_I (multiplied by 10^3) are given in Appendix 2, calculated depending on x^*, z_v for gearing with basic rack profiles RGU-5 and according to standards GOST 15023-76 and GOST 30224-96.

Table 8. Coefficients of polynomial for determining K_A, K_P, K_F, K_I.

A	K_A		K_P		K_F		K_I	
A_{00}	0.	500451	0.	504415	0.	671719	0.	582345
A_{01}	-0.	000028	0.	002086	0.	001965	0.	000354
A_{02}	-0.	000226	0.	000243	-0.	003341	-0.	000578
A_{10}	-3.	525752	-44.	858349	4.	219148	6.	234084
A_{11}	7.	070166	75.	350025	-16.	603538	-10.	257295
A_{12}	-3.	209129	-41.	925433	19.	399908	5.	492029
A_{13}	-0.	197789	11.	149256	-11.	344788	-1.	563736
A_{14}	0.	031285	-2.	169970	2.	756522	0.	273820
A_{20}	-32.	805630	2129.	902269	610.	883607	10.	822616
A_{21}	103.	304004	-6044.	771139	-1833.	839977	-14.	946592
A_{22}	-118.	497799	6424.	751507	2019.	196384	6.	011590
A_{23}	58.	909423	-3030.	701742	-972.	238396	-0.	600651
A_{24}	-10.	728590	535.	424031	173.	565337	-0.	033691
A_{30}	-79.	121131	-6527.	637913	-3427.	396257	-655.	088458
A_{31}	119.	327471	19110.	756242	11276.	993105	1807.	712605
A_{32}	-20.	094216	-20902.	274344	-13463.	841058	-1896.	516460
A_{33}	-37.	956701	10120.	545289	6977.	445641	892.	547177
A_{34}	14.	748268	-1830.	565069	-1332.	644440	-158.	039576

Chapter 4

Control of Tooth Cutting of Cylindrical Novikov Gearing

A number of methods of checking the correctness of Novikov gear tooth cutting is known from [44, 86, 197]: by means of balls, edge gear tooth snap gauges and gear tooth comparators, along the length of the common normal, and others.

Direct control of the hob penetration depth can be performed, for instance, by the accurate setting of the hob with respect to the workpiece, but this method is inconvenient. The most convenient and applicable practice methods are measurement of chordal dimensions of a tooth (tooth thickness along the chord and its height up to the chord) and according to the length of the common normal [98].

4.1. Control of Tooth Chordal Dimensions

This method is performed by means of the edge gear tooth snap gauge and is usually applied for large-dimension gears. Its drawback is in the necessity to locate the measuring tool (edge gear tooth snap gauge) on the outer diameter of the measured gear, which can differ from the nominal one In this connection, corrections must be introduced by the design value of the chordal height, and the measurement accuracy is reduced.

Chordal dimensions (thickness and height) of a tooth are determined by coordinates of the normal section of the tooth addendum surface. Investigations showed that these coordinates coincide to high precision with corresponding coordinates of the face profile of an equivalent straight-tooth gear In order to determine them, it is necessary to solve the enveloping task in a planar engagement.

In some system $\widetilde{S}_p$, related to the rack, the tooth profile of the latter can be expressed by the following parametric equations:

$$\begin{cases} \widetilde{x}_p^* = \rho_a^* \sin\vartheta + \left(x^* - x_a^*\right) \\ \widetilde{y}_p^* = -\rho_a^* \cos\vartheta + l_a^*, \end{cases} \tag{4.1}$$

where x^* is the shift coefficient of the basic rack profile when cutting the gear, ϑ is the current angle of the addendum profile. In the system $\widetilde{S}_p$, the axis $\widetilde{x}_p$ is directed upwards along the axis of tooth symmetry, the axis $\widetilde{y}_p$ is directed to the right along the pitch line.

Let's choose two more systems – the fixed system $\widetilde{S}_0$ with the reference point in the pitch point and the moving system $\widetilde{S}$, related to the rotating gear and which has the reference point in the center of its rotation. Using the transition matrix from the moving system $\widetilde{S}_p$ to the system $\widetilde{S}$ in the form [146]

$$M = \begin{Vmatrix} \cos\varphi & -\sin\varphi & 0.5z_v(\cos\varphi + \varphi\sin\varphi) \\ \sin\varphi & \cos\varphi & 0.5z_v(\sin\varphi - \varphi\cos\varphi) \\ 0 & 0 & 1 \end{Vmatrix}, \tag{4.2}$$

we obtain the equation of the gear tooth profile in the system $\widetilde{S}$ with the reference point, transferred to the pitch point:

$$\begin{cases} \widetilde{x}^* = \widetilde{x}_p^* \cos\varphi - \widetilde{y}_p^* \sin\varphi + 0.5z_v(\cos\varphi + \varphi\sin\varphi - 1), \\ \widetilde{y}^* = \widetilde{x}_p^* \sin\varphi + \widetilde{y}_p^* \cos\varphi + 0.5z_v(\sin\varphi - \varphi\cos\varphi). \end{cases} \tag{4.3}$$

Here, φ is the angle of gear rotation (form-generating parameter), $0.5z_v$ is the reduced pitch radius of the gear.

Parameter φ can be determined according to the equation of the normal to conjugated profiles

$$\frac{X^* - \widetilde{x}_p^*}{e_x} = -\frac{Y^* - \widetilde{y}_p^*}{e_y}, \tag{4.4}$$

where X^*, Y^* are the coordinates of the pitch point in the system $\widetilde{S}_p$, e_x, e_y are projections of the unit normal vector on the coordinate axis.

Determining $e_x = \partial\widetilde{y}_p^* / \partial\vartheta$, $e_y = \partial\widetilde{x}_p^* / \partial\vartheta$ and taking $X^* = 0,\quad Y^* = 0.5z_v \cdot \varphi$ (parameter of the rectilinear rack motion), after the transformations we obtain:

$$\begin{cases} \widetilde{x}^* = A^* \cos\varphi + B^* \sin\varphi - 0.5z_v, \\ \widetilde{y}^* = B^* \cos\varphi - A^* \sin\varphi, \end{cases} \tag{4.5}$$

where A^*, B^*, φ coincide correspondingly with A_S^*, B_S^*, φ_S - see the paragraph 3.2 of the Chapter 3.

It is recommended to choose the angle α_k (Fig. 3-5) or the angle close to it as the parameter ϑ.

Now the measuring tooth height up to the chord is

$$h_x = m\left(h_a^* + x^* - \widetilde{x}^*\right) \tag{4.6}$$

and the measuring thickness along the chord is

$$S_x = 2m\widetilde{y}^*. \tag{4.7}$$

4.2. Control of the Tooth along the Length of the Common Normal

This method of control is the most convenient, quite accurate and widespread.

Strictly from mathematical point of view, the task of finding the common normal is reduced to recording the equations of normal lines to opposite tooth flanks and imposing the conditions, for which these normal lines will take the same location in space, and tangent planes in calculation points of opposite flank will be parallel. In principle, such task has been solved in [197], and in [129], working formulas are given to determine the length W of the common normal for gears with basic rack profiles, having the shift x_a of the addendum curvature center.

If $\alpha_{min}, \alpha_{max}$ denote minimum and maximum angles of the basic rack tooth profile (of the rack) within the height of operating measuring segment, then parametrical angles will be determined correspondingly

$$\varphi_{min} = arctan\left(tan\,\alpha_{min} / \cos\beta\right), \qquad \varphi_{max} = arctan\left(tan\,\alpha_{max} / \cos\beta\right). \tag{4.8}$$

In order to provide the reliability of measurement, it is reasonable to take the following values as angles α_{min}, α_{max}: $\alpha_{min} = 1.06\alpha_p$, $\alpha_{max} = 0.85\alpha_a$.

The range of measured common normal line lengths will be determined by the minimum and maximum numbers z_n of gear teeth, gripped by the snap gauge:

$$(z_n)_{min} = entier\left\{\frac{1}{\pi}\left[\varphi_{min} + 2l_a^* + \frac{2\left(x^* - x_a^*\right)}{\cos\beta\cdot tan\varphi_{min}} + \frac{z\sin(2\varphi_{min})tan^2\beta}{2}\right] + 1\right\} + 1, \tag{4.9}$$

$$(z_n)_{max} = entier\left\{\frac{1}{\pi}\left[\varphi_{max} + 2l_a^* + \frac{2\left(x^* - x_a^*\right)}{\cos\beta\cdot tan\varphi_{max}} + \frac{z\sin(2\varphi_{max})tan^2\beta}{2}\right]\right\} + 1. \tag{4.10}$$

If it turns out that $(z_n)_{min} > (z_n)_{max}$, the common normal can not exist.

The current parameter φ is determined according to the transcendental equation,

$$\varphi = \frac{\pi(z_n - 1) - 2l_a^*}{z} - \frac{2(x^* - x_a^*)}{z\cos\beta \cdot \tan\varphi} - \frac{\sin(2\varphi)\tan^2\beta}{2}. \tag{4.11}$$

The equation (4.11) is solved by a method of successive approximations. Setting the angle φ_0 in the first approximation for the right part of the equation, the value $\varphi_1 \neq \varphi_0$ is obtained for the left part, which is then substituted again into the right part, and so on until both parts of the equation coincide with the required accuracy ε_φ. Experience shows that for $\varepsilon_\varphi = 10^{-5,}$ it is enough to make 4-5 iterations at the most.

Any integer $(z_n)_{min} \le z_n \le (z_n)_{max}$ is substituted into the equation (4.11) as z_n.

Now the desired length of the common normal will be

$$W = 2m\left[\left(\frac{z\sin\varphi}{2\cos\beta} + \frac{x^* - x_a^*}{\sin\varphi}\right)\sqrt{1 + \cos^2\varphi \cdot \tan^2\beta} + \rho_a^*\right]. \tag{4.12}$$

For helical gears, to which Novikov gears are related, the possibility to measure the common normal is determined by the sufficiency of the toothing width b of the measured gear according to the condition:

$$W < b / \sin\beta. \tag{4.13}$$

Measurement of the value W is performed over surfaces of active addendums. When applying basic rack profiles of the second group, having the shift of dedendum profile from the axis of tooth symmetry and the shift of addendum profile to the opposite side, a “recess” is generated on the tooth in the place of conjugation of the active dedendum with the conversion segment. During the process of measuring the length of the common normal grip of the snap gauge can thrust into this “recess”, which complicates measurements. That is why it is recommended to choose such length W of the common normal, for which the noted phenomenon will be excluded.

The condition of trouble-free measurement of the length W of the common normal for the given number z_n of gripped teeth is

$$W \ge W_y, \tag{4.14}$$

where W_y is the length of the common normal, measured over the “recess” and determined according to (4.9) - (4.12) with substitution of ρ_a^* by ρ_q^*, x_a^* by x_q^* and l_a^* by $-l_q^*$ (see Fig. 4, 5).

If the condition (4.14) is not fulfilled for the whole gear, it is recommended to manufacture a special gauge for measuring the length of the common normal [125].

Such a gauge (Fig. 11) has a limiting plane *1,* which rests upon vertex edges of teeth when measuring with its go gauge and therefore, it does not let grips *2* reach the "recess" *3* and thrust to it (Fig. 11a).

As calculations show, the dimension t_g from grips ends to the limiting plane *1* should be designed within the range *(1.05...1.15)* h_a, and its reliability can be easily checked graphically for an equivalent spur gear.

A certain clearance Δ should be generated between the plane *1* and vertex edges of teeth (Fig. 11b) when measuring by a no-go gauge.

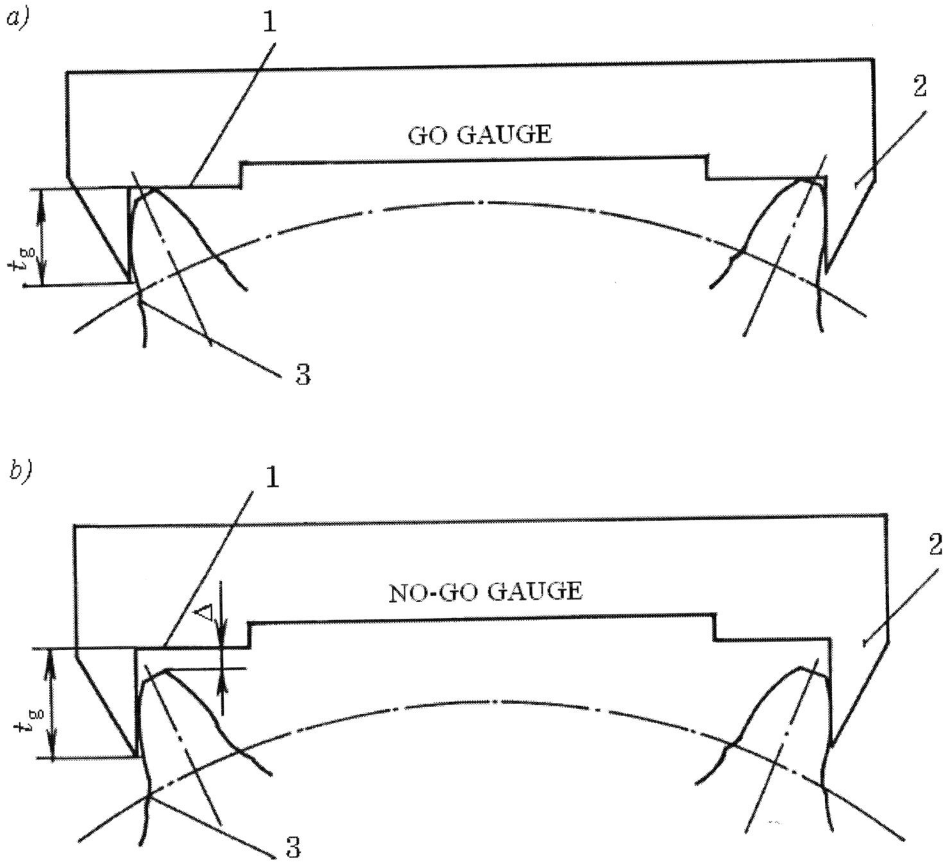

Figure 11. Scheme of common normal length measurement in the presence of "recess" on tooth surfaces: *a*) – by the go gauge; *b*) – by the no-go gauge.

To measure the length of the common normal for teeth, made according to the basic rack profile GOST 30224-96, the main active segment of the addendum should be used. But if it doesn't fulfill the condition (4.13)), then it is possible to apply the additional convex segment, taking $\alpha_{max}= 22^\circ$, $\alpha_{min}= 11^\circ$, assuming that $x_a^* = 0$ and substituting ρ_a^* by $\rho_a'^*$ and l_a^* by $l_a'^*$.

Chapter 5

MANUFACTURING ASPECT OF CYLINDRICAL NOVIKOV GEARING ADAPTABILITY

5.1. SOME GENERAL ASPECTS OF ACCURACY OF GEARING AND THE TOOTH MACHINING TOOL

Peculiarities of Novikov gearing tooth profiles lead to its greater sensitivity to radial errors, comparing with involute gearing, causing the shift of contact areas along the height, however, due to local character of contacting, NG is less sensitive to errors like misalignment, angular displacement of axes and so on, which cause load concentration along the tooth length in gearing with linear contact. Reduction of NG total sensitivity to various types of errors is considerably stipulated by a good run-in of contacting teeth, where the hardness of active flanks can be both high and low, leading to reduction of load concentration at certain segments of teeth and equalization of its distribution among contact areas. At present, there is no state standard on accuracy characteristics for NG, like for involute gearing [40].

In due time, recommendations on assigning the accuracy characteristics for NG with tooth hardness under *HB 350* and the basic rack profile like Ural-2N [85] were developed; which are still being applied by industrial enterprises in one way or another. Investigations of NG accuracy [188 and oth.], which had been carried out later, revealed the possibility to apply a number of manufacturing techniques, compensating the errors appearing in the gearing.

Along with the development of new progressive basic rack profiles and the design of NG with surface hardened teeth, the necessity arose to develop a standard on accuracy characteristics and methods of NG control, which could take into account its specific character. Activity in this field has been rather intensively performed [48-50], but unfortunately, it is still not finalized as the ultimate edition for the Russian standard. In this connection, further discussion and calculations have to be based on the existing standard [40] for involute gearing, especially because there is no principal difference between these two types of gearing according to kinds of accuracy characteristics (kinematic accuracy, smoothness, tooth contact and oth.) and methods of certain parameters control (gear rim radial run-out, variation of the common normal length, difference of circular pitches, deviation of the axial pitch along the normal line.).

In analyzing and summarizing the production experience of NG accumulated during the years, let's note several peculiarities, which are, in our opinion, should be taken into account.
1. It is known that, in large-scale production of involute gearing, a twin-flank method of control [215] is widely used that allows to fix variations of the testing center distance per one revolution of the gear, "jump" at one tooth, and also to determine the absolute deviation of the center distance from the design value. Such verification is performed by means of a precisely manufactured testing gear.

The twin-flank control is very useful for NG, since it allows the prompt obtaining of the important information on deviations of the center distance and its variations, which are the part of the total radial error, causing the contact pattern deviation along the height in operating gearing. By means of the twin-flank control, it is possible, in particular, to estimate the heat treatment influence on the radial error variation ("swelling" of the gear after chemically heat treatment and so on), performing the control before and after the heat treatment.

Production of a precise, testing Novikov gear (4-5th degree of accuracy) represents certain manufacturing difficulties. However, the way to overcome this situation has been proposed, by implying the proven possibility to use a precise helical involute gear with a specially calculated angle of basic rack profile [157] as the testing one, which makes the implementation of the twin-flank control for Novikov gears very prospective in large-scale production.

2. Tooth profile of hobs, intended for manufacture of Novikov gears, is usually measured at the projection apparatus according to precisely produced Plexiglas samples, the profile deviation is regulated by the standard [46], depending on the hob accuracy rating.

However, experience showed that, besides the general control of the hob tooth profile, it is necessary to control the profile deviation at points of convex and concave active areas, corresponding to theoretical points of contact on the tooth of the basic rack profile (Fig. 12), by means of a toolmaker's microscope. The deviation ΔS must be under the half of the tolerance f_{f0} for deviation of the hob profile on active areas of addendum and dedendum.

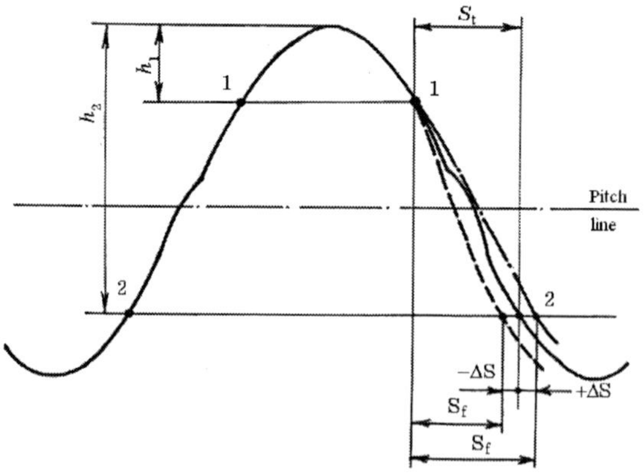

Figure 12. Deviation $\pm\Delta S$ of the actual dimension S_f from the theoretical value S_t between nominal points *1* and *2* of the hob tooth profile, measured separately at left and right flanks.

The necessity of the described regulations is caused by the requirement to provide a more reliable operation of gearing both lines of action with account of manufacturing conditions, under which the pinion and gear of the pair are not usually cut by one and the same specimen of the hob.

Verification of the deviation $\pm\Delta S$ is provided in the developed with assistance of the authors standard RD2 I41-16-88 on hobs for Novikov gears with the basic rack profile according to the standard GOST 30224-96.

3. Manufacturing radial error Δr depends mainly on three components: radial run-out F_{rr} of the pinion and gear, center distance deviation f_{ar} according to the casing borings and additional shift E_{Hr} of the basic rack profile when producing the pinion and gear, that is,

$$\Delta r = f\left(F_{rr1}, F_{rr2}, f_{ar}, E_{Hr1}, E_{Hr2}\right). \tag{5.1}$$

As it is known, the additional shift of the basic rack profile in the involute gearing is used to provide a guaranteed backlash in order to prevent the jamming of the pair when operating under overheating and so on.

Applying all the enumerated in (5.1) values of deviations [40] for NG, the radial error Δr will exceed the ability of the gearing to "keep" the contact area within the limits of tooth flanks according to the height and will lead to abrupt performance degradation.

If the values F_{rr} and f_{ar} depend mainly on the state of tooth-cutting equipment, predetermined by a technological production level and they are regulated with difficulties, the value E_{Hr} is, on the contrary, easier to be controlled.

For this reason, in order to reduce the value Δr, it is necessary to decrease the basic rack profile shift tolerance $\widetilde{T}_H$, using tolerances F_r and f_a [40], especially because the nominal backlash j_k for NG is usually provided by configuration of the basic rack profile.

In further calculations, we'll assign the tolerances F_r, f_a according to [40], and the tolerance for the basic rack profile shift will be calculated by the following relation, obtained according to the analysis of manufacturing process of Novikov gears in large-scale production:

$$\widetilde{T}_{H\Sigma} = 0.685[0.06 + 0.012(k-3)]m^{0.33}, \tag{5.2}$$

where $\widetilde{T}_{H\Sigma}$ is the total tolerance for the basic rack profile shift, which will be further understood as the summarized tolerance of the pinion and gear for the hob penetration depth when hobbing; k is the number of the degree of kinematical accuracy according to [40].

Distribution of the tolerance $\widetilde{T}_{H\Sigma}$ between the pinion and gear can be determined according to relations:

$$\widetilde{T}_{H1} = \widetilde{T}_{H\Sigma} / (1 + u^{0.2}); \qquad \widetilde{T}_{H2} = \widetilde{T}_{H\Sigma} - \widetilde{T}_{H1}. \tag{5.3}$$

5.2. Reduced Manufacturing Radial Error of Gears Production and Assembly

As applied to large-scale or mass production of gears, the reduced manufacturing radial error Δr must be considered in the stochastic aspect [197]:

$$\Delta r = \gamma_c (R_c)^{0.5} / K'_c \, , \tag{5.4}$$

where γ_c is the coefficient, characterizing the assigned probability of contact pattern overrun beyond the limits of the tooth active flanks, when the gearing workability is worsened; R_c is the total quadratic deviation, calculated as

$$R_c = F_{r1}^2 + F_{r2}^2 + \widetilde{T}_{H1}^2 + \widetilde{T}_{H2}^2 + f_a^2 \, ; \tag{5.5}$$

K'_c is the coefficient, depending on the law of errors distribution, for normal distribution $K'_c = 3$. Taking into account that the law of distribution of certain components of (5.5) is unknown, it should be carefully taken in calculations, that $K'_c = 2.3...2.5$.

In its turn, the coefficient γ_c depends on the accepted "risk level" t_c. On the basis of [9], let's present this relation in the form of a table (Table 9).

Table 9. The dependence of the coefficient γ_c on the "risk level" t_c.

t_c, %	0.001	0.01	0.05	0.1	0.5	1	3	4	5	6	7	10	12	20
γ_c	4.42	3.89	3.48	3.29	2.81	2.58	2.17	2.05	1.96	1.88	1.81	1.65	1.56	1.28

For more precise calculations, the following formula [9] should be applied:

$$\gamma_c = \omega - \frac{\sum_{i=0}^{2} c_i \omega^i}{\sum_{i=0}^{3} d_i \omega^i} \, , \tag{5.6}$$

where $\omega = \sqrt{ln(200/t_c)^2}$, $c_0 = 2.515517$, $c_1 = 0.802853$, $c_2 = 0.010328$, $d_0 = 1$, $d_1 = 1.432788$, $d_2 = 0.189269$, $d_3 = 0.001308$.

For practical calculations of general engineering gearing, it is recommended to take $t_c = (3...5)\%$.

In practice, gears are often manufactured in small amounts, down to single specimens, for example, for testing and so on.

In such cases, the law of large numbers stops its functioning, the term "risk level" losses its essence, since the risk must be excluded ($t_c=0$) and the manufacturing error Δr for small amounts of manufactured pairs is equal to the sum of tolerances

$$\Delta r = F_{r1} + F_{r2} + \widetilde{T}_{H\Sigma}. \tag{5.7}$$

The deviation f_{ar} of the center distance can be considered in this case not as a random quantity, but as a systematic error, which should be taken into account by a direct measurement according to casing borings.

5.3. Manufacturing Adaptability of Gearing

It is known that radial errors of gearing manufacture and assembly cause the height shift of the contact pattern from the nominal position towards the tip edges of teeth or the conversion near-pitch-point segment.

Fig. 13 shows the tooth *1* of a gear, a contact pattern *2* on its surface is shown in the nominal position (if radial errors are absent) and in the shifted position *3* near the tip edge *4* of the tooth under the action of radial errors. In the latter case, a part of the contact pattern may shift beyond the limits of the tooth active flank, resulting in the increased contact stresses (the so-called "boundary effects" will appear) and the workability of gearing will be worsened.

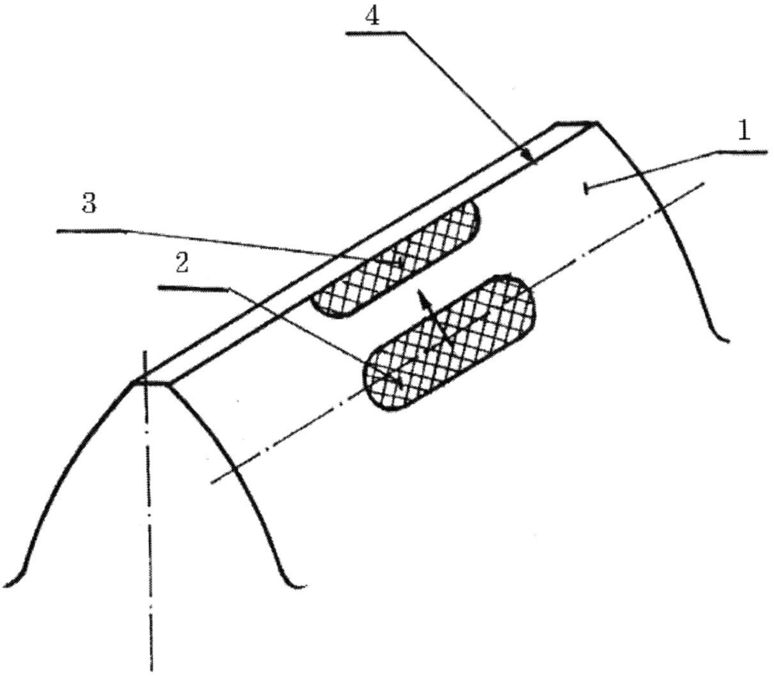

Figure 13. Scheme of height shift of the contact pattern on the tooth flank under the action of radial errors.

The adaptability of a gearing means its ability to "keep" the contact pattern within the given tooth height limits without malfunctions. In other words, it is the degree of gearing sensitivity to radial errors of gears manufacture and assembly.

If the radial error, which is allowable for a gearing without performance degradation (taken error) is denoted as Δa, then, obviously, in order to provide the enough adaptability, the following condition should be fulfilled

$$\Delta a \geq \Delta r\,. \tag{5.8}$$

Violation of the condition (5.8) results in the situation when a certain part t of the half-width (minor semi axis) b_H of the elliptical contact pattern runs beyond the limits of the tooth active flank with the noted above consequences.

On the basis of [179, 197], the parameter Δa can be presented as follows $\Delta a = \Delta\rho(sin\alpha_a - sin\alpha_p)$.

In its turn, the parameter Δa is equal to the sum of two components

$$\Delta a = \Delta a_a + \Delta a_p\,, \tag{5.9}$$

where $\Delta a_a = \Delta\rho(sin\alpha_a - sin\alpha_k)$ is the error, taken by segments of addendums and dedendums from the nominal point of contact (with design angle α_k) towards the increase of profile angles;

$\Delta a_p = \Delta\rho(sin\alpha_k - sin\alpha_p)$ is the error, taken by segments of addendums and dedendums from the nominal point of contact towards the decrease of profile angles.

The shift of contact pattern towards the increase of profile angles is limited by three factors: the height of the segment itself, the inadmissible reduction of the backlash j_w and the limiting minimum radial clearance c.

Considering the first factor, one can write:

$$\Delta a_{ab} = \Delta\rho[sin(\alpha_a - b_H / \rho_a) - sin\alpha_k]\ K_l\,. \tag{5.10}$$

If it is accepted, that the backlash j_w must not become less than the minimum guaranteed backlash j_{nmin}, regulated, for instance, according to [40], then one can write

$$\Delta j_w \leq j_w - j_{nmin}. \tag{5.11}$$

In turn, when increasing the profile angle from the value α_k up to a certain value α_{aj}, the reduction Δj_w of the backlash can be determined as $\Delta j_w = 2\Delta\rho(cos\alpha_k - cos\alpha_{aj})$, whence

$$\alpha_{aj} = arccos[cos\alpha_k - \Delta j_w /(2\Delta\rho)]\,, \tag{5.12}$$

and then

$$\Delta a_{aj} = \Delta\rho(\sin\alpha_{aj} - \sin\alpha_k). \tag{5.13}$$

Having accepted the minimum allowable value of the radial clearance $c_{min} = j_{nmin}$, letus write the corresponding restriction for the shift of the contact pattern

$$\Delta a_{ac} \le c - j_{nmin}. \tag{5.14}$$

On the basis of (5.10), (5.13) and (5.14), the minimum value should be taken as

$$\Delta a_a = \min\{\Delta a_{ab},\ \Delta a_{aj},\ \Delta a_{ac}\}. \tag{5.15}$$

As for the second component of (5.9), that is, the parameter Δa_p, it can be limited only by the necessity to locate the half-width b_H of the contact pattern on the tooth flank:

$$\Delta a_p = \Delta\rho[\sin\alpha_k - \sin(\alpha_p + b_H / \rho_a)] \cdot K_l. \tag{5.16}$$

For the basic rack profile according to the standard GOST 30224-96, the value α_p is taken in accordance with (3.4).

Calculating the shift Δa^0 of the theoretical contact point, taken by the gearing (that is, for $b_H = 0$), according to (5.9), (5.15) and (5.16), we'll obtain the relation to determine the part t of the half-width of the contact pattern, running beyond the limits of the active height of the tooth flank [110]:

$$t = (\Delta r - \Delta a)/(\Delta a^0 - \Delta a). \tag{5.17}$$

5.4. Tolerances for Tooth Thickness along Chord and Length of Common Normal when Tooth Cutting

Tolerance limits and ranges for the tooth thickness along the chord and length of the common normal depend on tolerance limits and ranges for the hob penetration depth (that is, for the shift of the basic rack profile).

In order to set the lower limit of the hob penetration depth tolerance, the fact established by the long-term practice should be taken into account, that the volume of the gear "body" (its "swelling") is increased as a result of heat treatment (especially a chemical heat treatment), causing the approach of contacting teeth profiles of a pair in action up to jamming.

In order to compensate this phenomenon, if the teeth of gears are not subjected to finishing (for example, tooth grinding), it is necessary to assign a certain value E_{HS} of the least basic rack profile shift when cutting, that is, a certain value of the additional hob penetration into the "body" of the workpiece, which can be taken roughly according to the relation [129]:

$$E_{HS} = 0.025 d^{0.17}, \tag{5.18}$$

where d is the pitch diameter of the gear.

For gears, which are not subjected to heat treatment or which are subjected to heat treatment before tooth cutting or tooth grinding, the value E_{HS} can be taken $2...2.5$ times less than the calculated one according to (5.18) or it can even not be taken into account ($E_{HS}=0$).

On the basis of the foresaid statements, the following relation for the lower limit $\tilde{T}_{\underline{H}}$ of the extreme deviation of the hob penetration depth when tooth machining can be written on the assumption of the symmetrical height location of the tolerance range on the active tooth flank [129]:

$$\tilde{T}_{\underline{H}} = \tilde{T}_H (0.5\Delta a - \Delta a_a) / \tilde{T}_{H\Sigma} - 0.5\tilde{T}_H + E_{HS}. \tag{5.19}$$

The upper limit $\tilde{T}_{\overline{H}}$ of the extreme deviation of the hob penetration depth will obviously be:

$$\tilde{T}_{\overline{H}} = \tilde{T}_{\underline{H}} + \tilde{T}_H. \tag{5.20}$$

Substituting the value $x^* - \tilde{T}_{\underline{H}} / m$, instead of the value x^* into formulas (4.5), (4.6), (4.7), we'll obtain the upper limit $\overline{S}_x$ of the tooth thickness along the chord for the chordal testing height $\overline{h}_x$, and when substituting $x^* - \tilde{T}_{\overline{H}} / m$, we'll obtain the lower limit $\underline{S}_x$ for the height $\underline{h}_x$. Then the upper $\tilde{T}_{\overline{S}}$ and lower $\tilde{T}_{\underline{S}}$ limits of extreme deviations of the testing chordal tooth thickness, denoted in detail drawings, and also the corresponding tolerance $\tilde{T}_S$ will be determined as

$$\tilde{T}_{\overline{S}} = \overline{S}_x - S_x; \tag{5.21}$$

$$\tilde{T}_{\underline{S}} = \underline{S}_x - S_x; \tag{5.22}$$

$$\tilde{T}_S = \tilde{T}_{\overline{S}} - \tilde{T}_{\underline{S}}. \tag{5.23}$$

Substituting the values $x^* - \tilde{T}_{\underline{H}} / m$ and $x^* - \tilde{T}_{\overline{H}} / m$ instead of x^* into formulas (4.11) and (4.12), we'll obtain the upper $\overline{W}$ and lower $\underline{W}$ extreme values of the common normal, correspondingly. Then the upper $\tilde{T}_{\overline{W}}$ and lower $\tilde{T}_{\underline{W}}$ limits of extreme deviations of the length of common normal, denoted in detail drawings, and the corresponding tolerance will be determined as

$$\widetilde{T}_{\overline{W}} = \overline{W} - W; \tag{5.24}$$

$$\widetilde{T}_{\underline{W}} = \underline{W} - W; \tag{5.25}$$

$$\widetilde{T}_{W} = \widetilde{T}_{\overline{W}} - \widetilde{T}_{\underline{W}}. \tag{5.26}$$

In conclusion, note that the value of the tolerance range (5.26) depends not only on $\widetilde{T}_H$, but also on the nominal value W of the common normal length, increasing with its growth.

Hence, a practical recommendation is made: maximum possible values should be chosen from the number of normal lines. It will facilitate the control and at the same time it will increase the possibility of free passing of the snap gauge grips over the "recess" of the tooth, that is, the condition (4.14) will be fulfilled.

Chapter 6

SPECIAL TYPES OF CYLINDRICAL NOVIKOV GEARING

Having developed a general method of gearing synthesis, M.L. Novikov proposed on its basis, the common contact point simplest law of motion in a stationary space – with the constant speed along the straight line (line of action), parallel to the pitch point line [158]. Such gearing, as it is known, is principally helical, that only operates theoretically due to the teeth axial overlap, and the load transmission is permanently accompanied by the appearance of the axial component F_x of the force in gearing. In cases of the limited axial overall dimension of the design, that causes the necessity to assign the increased values of tooth helix angles, axial forces may reach considerable values, acting destructively on bearing supports.

There are units where it is impossible to increase the load-bearing capacity by application of a conventional helical NG due to axial forces peculiar to it. One of the typical representatives of such units is the final (hub) gear drive of some models of tractors, containing a spur involute pair of gears. The peculiarity of these units is, on one hand, the provided possibility of only axial assemble-disassemble of a pair, predetermining also the design of the external bearing support of the driving pinion, capable to take only insignificant axial loads. On the other hand , the limitation of the axial overall dimension of the unit [108].

On the basis of the M.L. Novikov method in the Southern Federal University (SFU), new varieties of space cylindrical gearing were developed, they combine the load-bearing capacity at the level of NG with the abruptly reduced (down to zero) axial component of forces [238].

The concept of engagement of such gearing is shown in Fig. 14.

Letus choose a straight line as the line of action, that is inclined to the pitch point line, and let's direct the generating lines of the basic rack teeth cylindrical surface (and, therefore, the straight contact line K_1K_2 of the rack) under a certain angle Θ to the pitch plane parallel to the axial section of the rack.

When displacing the rack along the arrow A, the common contact point of flanks of the rack and the gear, engaging with it, moves along contact lines of both elements of the pair and along the line of action. Here, as evident from Fig. 14, the value of the contact normal segment to the face profile from the contact point to the pitch point line PP uniformly varies from the value l_{k1} (at the initial contact point K_1) to the value l_{k2} (at the final contact point K_2). The line of action (dotted line) in this case will be inclined, so that its projection on the gearing reference plane could make a certain angle β_l with the pitch point line. The following relation will be valid here

$$tan\Theta = tan\alpha_{kt}\, tan\beta_l\,, \tag{6.1}$$

where α_{kt} is the face angle of the rack tooth profile in the contact point.

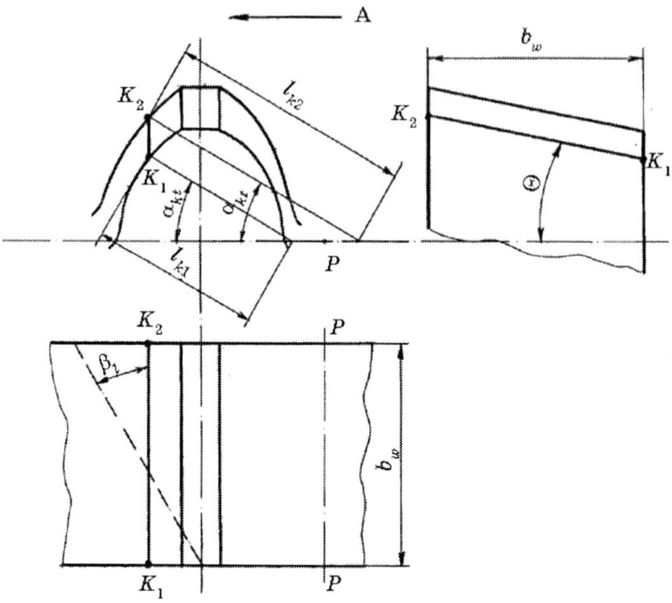

Figure 14. Addendum of the basic rack of the cylindrical Novikov gearing with straightened teeth in three projections.

Then the axial contact ratio for each (of two) line of action will be expressed by the relation:

$$\varepsilon_\beta = b_w\, tan\beta_l\, /\, p_t = (l_{k2} - l_{k1}) cos\alpha_{kt}\, /\, p_t, \tag{6.2}$$

where p_t is face (tangential) pitch of rack teeth.

According to relations of normal lines projections to contacting surfaces at the contact point [126], the relation between axial F_x and tangential F_t forces is easily determined:

$$F_x = F_t\, tan\alpha_{kt}\, tan\Theta\,. \tag{6.3}$$

Taking into account that, for a conventional NG

$$F_x = F_t\, tan\beta\,, \tag{6.4}$$

$$\varepsilon_\beta = b_w\, tan\beta /\, p_t \tag{6.5}$$

and comparing (6.1) - (6.5), we will obtain, that for similar b_w and ε_β the value F_x of the axial force of the considered gearing (letus call it NG with straightened teeth) is $cot^2\,\alpha_{kt}$ times less, than of the corresponding conventional helical NG. For the angle $\alpha_{kt} = 12^\circ...30^\circ$, we obtain the reduction of the axial force down to $3...22$ times.

Investigations and calculations, carried out for NG TLA with straightened teeth, showed that the value of angle β_l, necessary for normal operation, varies within the limits $10^\circ...20^\circ$, and the angle Θ is within the limits $3^\circ...6^\circ$ according to (6.1). This means, that the ratio F_x / F_t does not exceed $0.05...0.06$, moreover, the direction of action of the force F_x is not changed when the rotation is reversed, which turns to be quite valuable in practice.

It also results from the given formulas, that the gearing with straightened teeth has the operating width of toothing, that is, the axial overall dimension, also $cot^2\,\alpha_{kt}$ times less, than in the conventional gearing for the same ε_β and F_x.

Tooth cutting of gears of the considered gearing does not practically differ from tooth cutting of common shaping cutters and it is carried out at batch gear-milling semi-automatic machines (for example, of 5K32 model), either by a special taper mill [97], or by means of an additional adjustment gear train, connecting the vertical displacement of the carriage with the mill and the horizontal displacement of the table with the workpiece [133]. In the second case, the resulting feed of the tool with respect to the workpiece must be under the assigned angle Θ to the axis of the latter, in order to reproduce the tooth surface of the rack-type tool (Fig.14) in engagement with the machined gear. The root surface of the cut gears is conically shaped here with the angle 2Θ at the cone vertex. In order to create the uniform radial clearance, the surface of tooth tips is also made conical with the angle 2Θ at the cone vertex.

Fig. 15 schematically illustrates a pair of gears with straightened teeth, and Fig. 16 shows the photograph of such pair, the teeth of which are cut at the gear-milling machine 5K32, provided with the additional adjustment gear train.

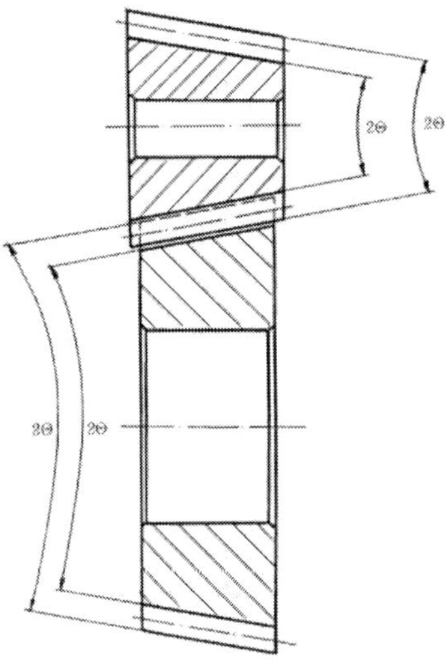

Figure 15. Schematic view of the cylindrical Novikov gear pair with straightened teeth.

Figure 16. Novikov gear pair with straightened teeth cut at the gear-milling machine 5K32.

A rather valuable quality of the considered gearing is the possibility of the engagement axial adjustment, which partially or completely compensates radial manufacturing errors. But even for the inaccurate axial mounting of the pair, the gearing conjugation is not broken [124].

In those cases, when the gearing must operate as an irreversible one, it can be designed so the operating load would be transmitted without the axial force ($F_x = 0$). For this purpose, it is necessary to arrange the basic rack contact line in such a way that its projection on the axial section of the rack would remain at the angle Θ, and on the pitch plane at the angle β to the pitch point line, determined from the relation [68]:

$$tan\beta = tan\alpha_{kt}\, tan\Theta . \tag{6.6}$$

According to the angle β, a differential change gear train of the gear-milling machine is adjusted.

Another interesting modification of the gearing with straightened teeth is the gearing with "pine-tree" teeth, shown in Fig. 17, the basic rack teeth contact line which is made in the form of a broken straight line [237]. Such gearing has the increased number of contact points in simultaneous engagement and, similar to common herring-bone gearing, it is capable to the axial self-aligning. Its advantage compared with a herring-bone gearing is that it is not required to reset a hob when manufacturing gears, owing to it there is no need for a groove, and the loss of processing time of tooth-machining is reduced, axial overall dimension of the gearing is decreased and the tooth becomes strong due to the unbroken (without groove) "arched" shape along the length. Moreover, the absence of the hob resetting automatically provides a strict coincidence of tooth symmetry planes for all segments of the gear rim, and this, in turn, leads to equalization of load distribution between segments and a more integrate application of the total operating length of teeth.

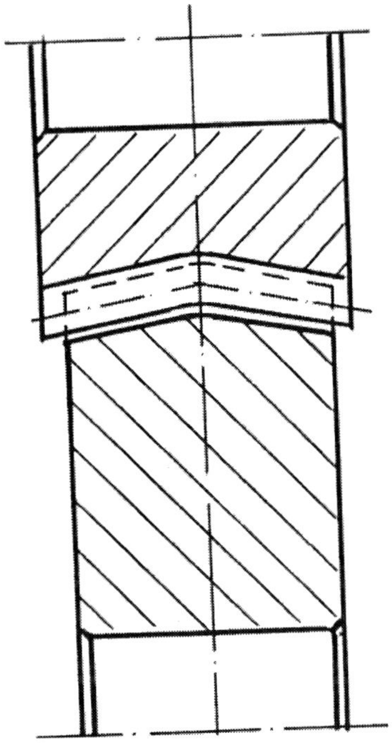

Figure 17. Novikov gearing with "pine-tree" teeth.

In order to check the working capacity of NG with straightened teeth, 16 sets of these pairs were made in SFU and for comparison, 16 batch produced (involute spur) pairs of the hub drive of the tractor T-4A, manufactured at the Altai Tractor Plant (ATP). The compared gear pairs were made according to the same technique of steel 20HN3A with nitrogen case-hardening and quenching up to *HRC (56...62)* and they had parameters: $m=8\ mm$, $z_1=13$, $z_2=57$, $b_w=70\ mm$.

When testing the prototype pairs with straightened teeth at test-rigs of ATP, it was established that the axial force in the engagement was about 4% of the tangential force and it did not cause (even under the accelerated modes of testing) premature failure or reduction of the scheduled durability of the single-row radial plain roller-bearings with short rollers and the end ring of 62000 series according to the standard GOST 8328-57 [108] mounted there.

Tests showed that the characteristic mode of failure of both prototype and batch produced gearing at the test-rig is the progressive pitting on flanks of the driving pinion with the subsequent chippage of the tooth along the generated concentrator. It was established preliminary [108, 126], that the operating time (in hours) before the failure for both usual and accelerated test modes of prototype pairs was, at the average, *2.3...2.4* times more than of batch produced pairs. That allows to hope for the effective application of NG with high-hardened straightened teeth instead of similar spur involute gearing in the units with restricted limiting requirements of the value and direction of axial forces and also of the axial overall dimension of the unit.

Chapter 7

DAMAGE VARIETIES OF NOVIKOV GEAR TEETH

The most frequent reason of failures of correctly designed, closed, well-lubricated Novikov gearing with surface hardness under *HB 350* is the damage of active tooth flanks, and for high surface hardness – tooth fracture. For gearing with the surface hardened layer of metal (nitration, nitro carburizing, carburizing), factors that determine the load-bearing capacity may become the surface failure of teeth or their fracture, caused by appearance of depth fatigue cracks.

Letus consider these damages in more detail. When tooth flanks are soft, the limited pitting is characteristic, observed during the initial period of operation [3]. As a rule, the limited pitting is related to load concentration on local segments, irregularities of flanks and appearance of fatigue cracks there. In gearing with the basic rack profile, according to the standard GOST 15023-76, the appearance of cavities on the involute segment within the near-pitch-point area is observed as the result of metal particles pitting. Furthermore, for soft, plastic flanks and the moderate load, cavities are rolled over, the load is redistributed more uniformly along the surface and the pitting is stopped. On the contrary, for hard and brittle flanks the pitting process is usually progressive. Cavities are considerably increased in dimensions due to metal chipping-off, and their number is increased. Tooth flanks become irregular, lubrication is polluted with pitting particles, which, in turn, intensifies wear. In most cases, gears are capable operating for some more time, before the started process of surface pitting leads to complete tooth destruction.

In gearing with the near-pitch-point area excluded from operation, (basic rack profiles RGU-5, Don-63 and other) the pitting takes place on active tooth flanks, but under loads many times exceeding those loads when pitting of the near-pitch-point area occurs.

Lubrication has a considerable influence on fatigue pitting development. If it is absent or it is limitedly supplied, greater loads are required for the pitting appearance than if the lubrication is enough. Nowadays, the well-founded accounting of lubrication influence on pitting is still impossible [43, 183] due to insufficient investigations of this problem.

In open gearing, the pitting is observed very rarely, since the thin surface layer is worn out quicker; then the metal "is fatigued".

Under the influence of cyclically varying contact stresses, the fatigue cracks can appear not only on the tooth flank, but also inside the metal [61, 196, 226]. The depth of their appearance depends on properties of tooth material, the value and distribution of acting and residual stresses. For surface hardening, the contact load, which is not dangerous for hard surface layers, can cause subsurface stresses, exceeding the endurance limit and leading to the appearance of depth fatigue cracks [31]. During the operational process of the gearing, depth cracks are developed and go out to the surface, and as the result, considerable areas of the surface are peeled off.

It should be kept in mind that depth failures can be caused by short-term overloads, accumulated during the continuous period of operation, which are difficult to account for. The transience and danger of fatigue cracks development process, leading to tooth fracture, make us seriously reckon the possibility of the appearance of this type of tooth failure of NG. In order to increase the reliability of operation of heavy-loaded gearing, it is necessary to reasonably assign the hardness of the tooth core and optimal depth of the hardened layer.

Another dangerous cause of gearing failure is fracture – a complete destruction of gear material, leading to the tooth chipping-off. Fractures after a single action of the load are observed comparatively rarely. Fatigue fractures under the action of variable stresses during a certain durability of the gear are the most characteristic. Fatigue fractures appear under stresses, less than the yield point of the tooth metal. The initiation of a fatigue crack is promoted by the presence of micro defects and load concentrators in the root and on the conversion concave segment of the tooth. The typical fatigue fracture is characterized by the presence of the failure site, the zone of a fatigue crack and the break-off area. Independently on the material viscosity, the fatigue fracture is of the brittle nature.

The process of the fatigue crack initiation begins, as a rule, in the zone of lower or upper stress concentrator on the tensile flank. Gradually, under the influence of a variable load, the microcrack is developed deep into the tooth body. The tooth is broken off – more frequently near the end faces, more seldom in the middle or along the whole gear rim [129, 172, 239]. The character of fracture is considerably influenced by the shape of the basic rack profile, the presence of load concentration due to manufacture and assembly errors and the appearance of the fatigue pitting on the flank.

In some cases, the durability of Novikov gears can be limited by mechanical wear, displayed as the result of friction interaction of tooth pairs. Depending on the gears operation conditions, the wear during the process of contact and rubbing of metallic pairs under the action of rolling and sliding friction and also the abrasive wear are distinguished. The presence of a considerable flanks roughness predetermines the run-in wear during the initial operation period, as the result of which the correct relation of pinion and gear tooth hardness, the flanks become so smooth that microroughnesses do not exceed the thickness of the oil film. After that, the wear intensity is considerably decreased.

In literature, the contradictory opinions about the wear resistance of NG are given [51]. Compared with the involute gearing, it has increased values of the reduced curvature radius and higher rolling speeds, promoting the increase of a continuous hydrodynamic oil layer thickness. Better conditions of the lubrication of contacting surfaces in NG compared with the involute one are not only justified theoretically, but also proved experimentally [83]. The thickness of the oil film of the gearing with the involute profile, measured according to the electrical resistance, was almost two times less than the thickness of the oil film for the similar NG. At the same time, certain testing [51] showed the absence of a non-wear mode of operation of NG, which exists in involute gearing. Experimental investigations discovered a considerable wear resistance of TLA gearing compared with the OLA one.

If abrasive particles (grains, metallic particles, abrasive grit) are present in oil, abrasive wear is observed. Its intensity is considerably determined by the hardness of active flanks and oil viscosity.

Under the action of significant overloads, distortion of the active flanks shape can take place as a result of plastic deformation. Teeth of low hardness are subjected to it to a greater extent [60]. As for teeth with hardened surface and soft core (carburized, nitrated and oth.), plastic deformation has often the character of local influxes in the form of ripples [240]. Such type of damage is characteristic for gearing, operating under heavy loads.

There are cases of scuffing of the large-pitch NG with insufficient difference of radii of the convex addendum and concave dedendum of the basic rack profile [47], and also of the gearing with the increased tooth height [14] with flanks hardness under *HB 350*.

Chapter 8

FATIGUE TESTING OF CYLINDRICAL NOVIKOV GEARING

During the period of preparation, and then the upgrading of the first standard for the basic rack profile of NG TLA, intensive tests of gear pairs with the basic rack profile Ural-2N, and then with the basic rack profile according to the standard GOST 15023-76, developed on its basis were carried out in 1960-1975 at the gearing laboratory of LMI under the leadership of Prof. V.N. Kudryavtsev [3].

The purposes of the tests were the definition of the NG load-bearing capacity and development of engineering techniques of its analysis.

The tested pairs had the tooth hardness $H_{HB} < HB\ 280$ and a comparatively low degree of accuracy, obtained after gear milling.

Altogether, 32 pairs with the module $m = 4\ mm$ of three dimension types were subjected to contact endurance tests: $z_1 = 29,\ z_2 = 30;\ z_1 = 12,\ z_2 = 59$ and $z_1 = 13,\ z_2 = 46$. Parameters of the first dimension type were as follows: $b_w = 36\ mm,\ \beta = 25^\circ$, the average hardness of flanks $H_{HB1} = HB\ 269,\ H_{HB2} = HB\ 255$, the tangential speed $v = 9.8\ m/s$, of the second dimension type: $b_w = 47\ mm,\ \beta = 18^\circ 47' 49''$, $H_{HB1} = HB\ 241,\ H_{HB2} = HB\ 207,\ v = 3.9\ m/s$.

When processing the results of the tests, the allowable experimental torques, according to the contact endurance T^E_{HP2} at the driven gear of the pair, were obtained, which proved to be equal to $702\ N \cdot m$ for the first batch, and $1005\ N \cdot m$ for the second one. Surface pitting almost always began in the zone of near-pitch-point involute segment It was distributing over active segments of the addendum and dedendum, yet it was not always progressive. The beginning of the pitting in the near-pitch-point area was the consequence of both gearing high sensitivity to radial errors of manufacture and assembly in the casing, that led to contact bands "compression" to the middle part of the tooth height, and of participation of the weak (regarding the contact) near-pitch-point involute area in operation.

Two groups of gears – *29/30* and *12/59* – were also subjected to fracture strength tests with performing 10 and 4 experiments correspondingly, as the result the allowable experimental torques according to the fracture strength T^E_{FP2} were obtained at the level $797\ N \cdot m$ and $\approx 1500\ N \cdot m$ correspondingly for the first and the second group of gears. It should be noted that the considerable scatter of testing results was observed, related evidently to high sensitivity of the gearing with the basic rack profile Ural-2N to errors. In spite of it, the authors of these tests made a conclusion on the rather high contact strength of NG with the basic rack profile Ural-2N, which was *2...2.5* times more than this type of strength for involute analogs, and on its fracture strength, corresponding to the level of the standard involute gearing.

The values of experimental allowable torques according to both contact and bending endurance are well agreed with the proposed [3] calculation relations, by authors since the latter are completely based on results of the carried- out tests.

The increased sensitivity to errors of gearing with the basic rack profile Ural-2N was especially visible during the tests, performed at LMI, among 9 surface hardened pairs with unground teeth (that is, with the low degree of accuracy), having the flank hardness $H_{HRC} = HRC\ 43$ [177]. The scatter of results was so significant that it was impossible to establish the allowable torques with enough reliability.

The analysis of the results led to the conclusion on irrationality of basic rack profile application according to the standard GOST 15023-76 for gearing with low accuracy with hard surfaces of teeth. As for accurate gearing, high contact endurance can be expected in this case for the gearing with the basic rack profile according to the standard GOST 15023-76 only until the involute near-pitch-point area starts operating.

At ZAFEA, the tests of high-precision gearing with ground flanks and basic rack profile DLZ-0.7-0.15 [172] were carried out.

This basic rack profile has the decreased tooth height and, similar to the basic rack profile according to the standard GOST 15023-76, it predetermines the increased sensitivity of the gearing to radial manufacture errors. However, if precisely produced, teeth with the given basic rack profile have the high fracture strength. This illustrates the results of the performed tests of pairs made of steel 12H2N4A with carburized hardened (up to *HRC 62*) and ground teeth, which showed the extreme experimental torque according to fracture endurance $T^{E}_{F\,lim\,2} = 3768\ N \cdot m$, exceeding *1.6* times similar torque for analogous high-hardened involute gearing with ground teeth [172].

Gearing with carburized teeth with basic rack profiles. “Don” passed the tests at LMI (the basic rack profile Don-68 was used, which did not become widespread) and in Rostov-on-Don NIITM [196], where pairs with both one line of action (the basic rack profile OLZ-63) and two lines of action (the basic rack profile Don-63) were tested, intended for mounting in cars ZIL and trolleybuses ZIU-5 as the main gearing of rear axles.

During the process of investigations, the task was set not only to replace involute pairs with ground teeth by Novikov pairs with unground teeth, but also to compare certain versions of basic rack profiles of the group “Don” between each other in order to refine their parameters. In particular, the low load-bearing capacity (due to edge contact) was established for gearing with basic rack profiles, having a small value (about 0.1) of the concave and convex radii difference $\Delta\rho^*$, which was the reason of further increase of $\Delta\rho^*$ up to the value *0.35*.

The performed tests of pairs with parameters $m = 8\ mm,\ z_1 = 10,\ z_2 = 32,\ b_w = 70\ mm,\ \beta = 28.13°$ with the basic rack profile Don-63, having the low sensitivity to errors

($\Delta\rho^* = 0.35$), showed their increased load-bearing capacity compared with involute ground pairs according to both contact and bending endurance, which was further proved by operation of these gears in trolleybuses [196].

An interesting result was obtained when testing the pair with the basic rack profile OLZ-63 – teeth of the pinion failed due to insufficient depth contact endurance under the small number of cycles ($N_1 = 3.2 \cdot 10^6$). Photographs (Fig. 18) show the appearance of the flute crack, which emerged later on the tooth surface. This experience stimulated the development of the calculation technique of NG with surface hardened teeth for depth contact endurance (see Chapter 12).

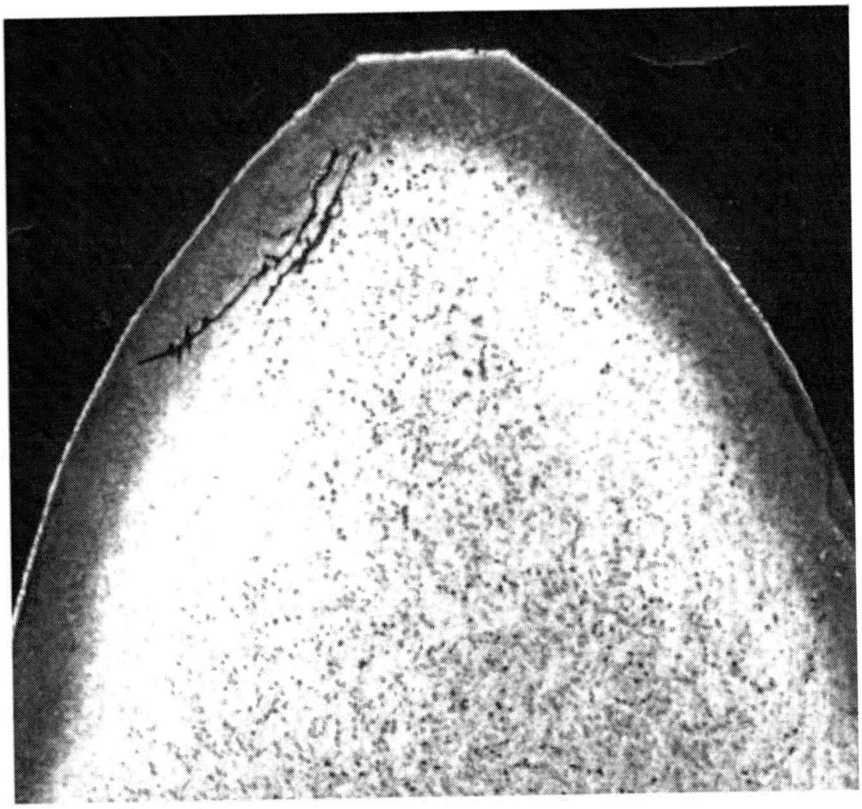

Figure 18. Microsection of the carburized tooth of the Novikov gearing pinion with the basic rack profile OLZ-63, with visible flute crack.

According to the decision of the Coordinating Board at USSR Minstankoprom on the problem of implementation of NG with hard teeth in this field, extensive tests of nitro carburized pairs with unground teeth made of steel 25HGM for general-purpose, mass - produced gearboxes were carried out at Izhevsk "OOO "Reduktor" and at VNIIReduktor (Kiev).

During the period of 1978-1984 at "OOO "Reduktor", two versions of gearing with low degree of accuracy (10-9-9B according to [40]) were tested – with the basic rack profile RGU-5A and RGU-5B, from which after analysis of results the basic rack profile RGU-5A was preferred, which was further called RGU-5.

Tests of pairs with parameters $m = 3.15\ mm,\ z_1 = 32,\ z_2 = 65,\ b_w = 45mm,\ \beta = 17.284^\circ,\ H_{HRC1} = H_{HRC2} = HRC\ 56$ [22] were carried out in casings of single-stage gearboxes TsU-160 (Fig. 19) under the tangential speed $v = 8.29\ m/s$ at test-rigs with the closed power flow, by loading with weights and with a circulating lubrication system with the oil MS-20. In all experiments, teeth were loaded up to failure of any element of the pair.

Figure 19. General view of a single-stage gearbox TsU-160; in the casing of which fatigue tests of nitro carburized Novikov pairs were carried out.

Tests were carried out at several levels of load counting on further correlation analysis of results. A batch of *13* pairs was selected for tests, subjected to the unified process of nitro carburizing with the subsequent quenching in the presence of "witnesses", showed the satisfactory structure of the core material and hardened layer according to the plant laboratory data, which predetermined further not a random (defective), but a fatigue character of tooth fracture.

After the short-term (*5-8* hours) rolling-in under the load $T_2 = 1000\ N \cdot m$, the tested pairs showed satisfactory arrangement of both contact bands along the height, testifying the rather quick run-in of even very hard flanks, and distribution of these bands along the whole teeth length.

The results of tests [22], grouped in ascending order of loads and operating time cycles, are shown in Table 10.

Photographs (Fig. 20 *a-f*) show the character of fracture in several experiments.

Table 10. Results of fracture endurance tests of nitro carburized pairs having unground teeth with the basic rack profile RGU-5.

Loading torque at the gear $T_2^E,\ N\cdot m$	Number of cycles before fracture N_{F1}, mln	Failed element of the pair	Character of tooth failure
3047	19.32	Gear	Tooth fracture at the addendum
	25.60	Pinion	
	33.84		
	37.25		Edge fracture of teeth at the dedendum
	52.52	Gear	Edge chippage of the tooth at the addendum
3352	5.79		Tooth fracture at the dedendum
3555	3.25		Edge chippage of the tooth at the addendum
	5.13	Pinion and gear	Tooth fracture at the addendum
	6.30	Pinion	Tooth fracture at the dedendum
	7.20		Tooth fracture at the addendum
	7.70	Gear	
	12.25	Pinion	Tooth fracture at the dedendum
4063	5.00	Pinion and gear	Tooth fracture at the dedendum (pinion) and addendum (gear)

Active flanks in all experiments were clean, bright, without any marks of contact damages. After *4...5* mln. cycles, the width of addendums and dedendums contact bands was practically stabilized, covering consequently about 80% of the active tooth height. This allows, in particular, to propose, that the torque $T_2^E = 3047\ N\cdot m$ (for $N_{F1} = 52.52\cdot 10^6$) is quite permissible according to contact endurance.

Correlation analysis of the relation between $\log T$ and $\log N$, performed according to techniques [1, 242] for small sample sizes, revealed the correlation factor *-0.8423,* and allowed to construct the regression line for the extreme and "safety limit" for the allowable experimental torques according to fracture with the probability of non-failure $P = 0.995$ [129, 239]. After the final processing of testing results, the exponent of the endurance curve at the sloping segment, characterized by the angle of regression line inclination, was obtained, equal to *8.273*, which is very close to the exponent $q_F = 9$ for involute gearing with hard teeth [43]. The sloping segment lasted approximately up to the number of cycles $N_{F1} = 30\cdot 10^6$, for which the allowable experimental fracture torque $T_{FP2}^E = 2181\ N\cdot m$ was determined. The obtained comparatively small difference between the extreme and allowable experimental torques showed acceptable for the statistical handling scatter of testing results, which is explained, to our opinion, by low sensitivity to errors of the gearing with the basic rack profile RGU-5.

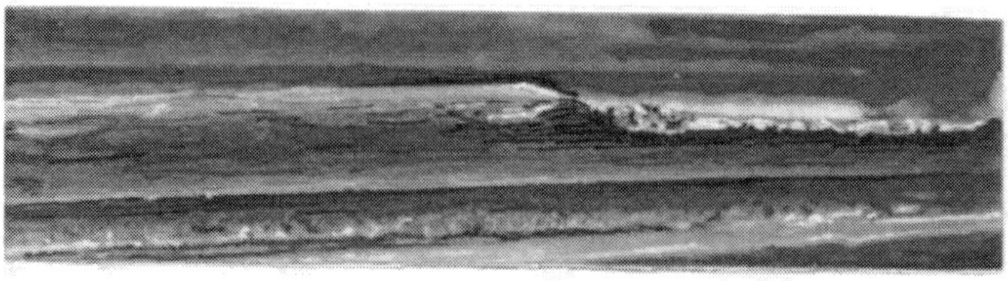

a) Pinion tooth breakage after $N_{F1} = 33.84 \cdot 10^6$ for $T_2^E = 3047\ N \cdot m$

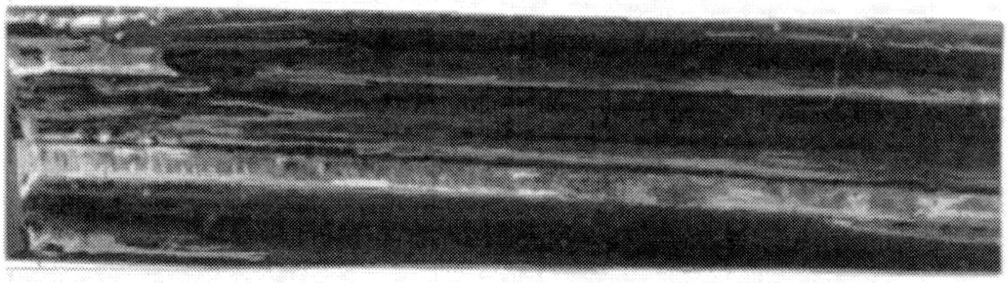

b) Gear tooth breakage after $N_{F1} = 52.52 \cdot 10^6$ for $T_2^E = 3047\ N \cdot m$

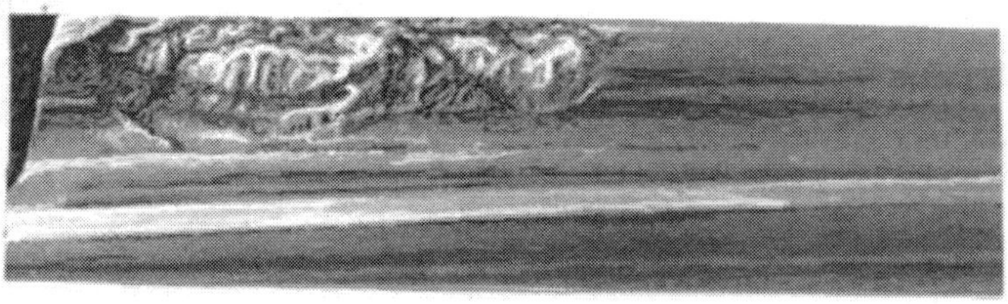

c) Pinion tooth breakage after $N_{F1} = 5.13 \cdot 10^6$ for $T_2^E = 3555\ N \cdot m$

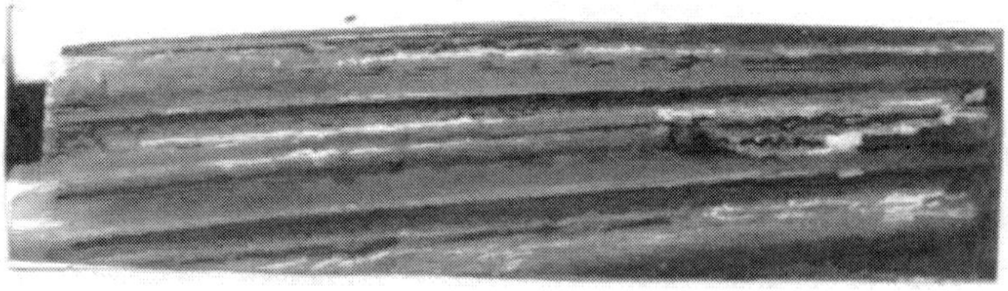

d) Pinion tooth breakage after $N_{F1} = 7.20 \cdot 10^6$ for $T_2^E = 3555\ N \cdot m$

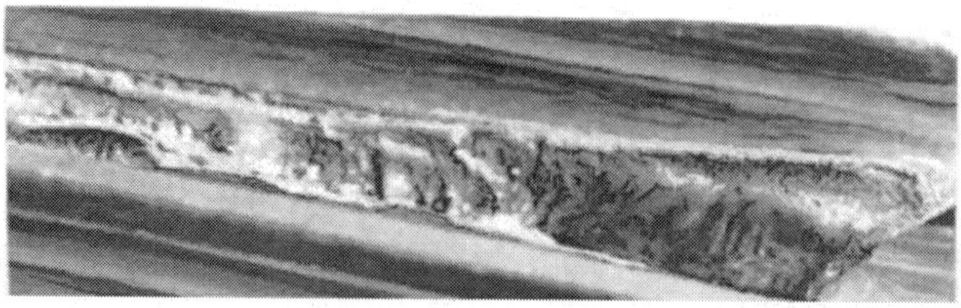

e) Pinion tooth breakage after $N_{F1} = 5.00 \cdot 10^6$ for $T_2^E = 4063\ N \cdot m$

Figure 20. Character of tooth breakage.

Long-term tests of similar nitro carburized involute pairs at "OOO "Reduktor" were accompanied by a considerable scatter of results, which caused the necessity to assign big design fracture safety factors and to therefore reduce the design allowable load. Researchers at LMI came to the same conclusion with respect to the involute gearing.

The results, obtained in the described series of tests, finally showed the increase of the load-bearing capacity according to fracture strength of NG with the basic rack profile RGU-5 *1.3…1.5* times [22] compared with involute analogues, which is explained to a great extent by better adaptability (thanks to the local, rather than the linear character of contacting) of NG to various misalignments in the engagement.

The pilot batch (600 items) of double-stage gearboxes Ts2U-160, produced at "OOO "Reduktor", both stages of which are provided with nitro carburized pairs with the basic rack profile RGU-5, have been in operation for many years without any complaint from consumers.

Summarizing the analysis of testing results of nitro carburized gearing with the basic rack profile RGU-5, note that probabilities of pinion and gear failure and also probabilities of tooth fracture in the zone of near-pitch-point concentrator (addendum) and in the fillet zone (dedendum) proved to be practically the same, indicating that relations of basic rack profile RGU-5 parameters are close to being optimal.

Further investigations led to the upgrading of the basic rack profile RGU-5, in order to increase the fracture strength due to the increase of the addendum strength without reduction of the dedendum strength. Based on the basic rack profile RGU-5, the Coordinating Board with the active assistance of authors of the monograph, developed the basic rack profile KS, standardized at first in the branch [177] for gearing with tooth hardness $H_{HRC} \geq HRC\ 35$ (with the possibility to be applied also for gearing with lower hardness), which then got the status of the Interstate Standard GOST 30224-96 [39].

At "OOO "Reduktor", comparative monitoring tests were carried out for nitro carburized ($HRC\ 56$) with the basic rack profile KS and involute pairs (5 experiments for each group), having the degree of accuracy 10-9-9 [40] and parameters: $m = 3.15mm,\ z_1 = 32,\ z_2 = 65,$ $b_w = 45\ mm,\ \beta = 17.284^\circ,\ a_w = 160\ mm.$

The results of these tests are shown in Table 11.

As it is evident from the Table, the involute gearing showed a considerably greater scatter of testing results, than the NG with the basic rack profile KS did. The extreme fracture torque for the latter at large number of cycles (not less than $N_{F1} = 30 \cdot 10^6$) proved to be approximately $T^E_{F\,lim\,2} = 3050\ N \cdot m$, which is *1.34* times more than for the involute analog.

The ratio between extreme (and, obviously, allowable) torques, according to fracture, reduced to one and the same number of cycles, for gearing with basic rack profiles KS and RGU-5 was *1.06*, which proves the correctness of theoretical background when developing the basic rack profile KS.

Table 11 also shows that, for gearing with the basic rack profile KS, the torque $T^E_2 = 3555\ N \cdot m$ is not less than the allowable one according to the contact endurance (for $N_1 = 51 \cdot 10^6$).

Table 11. Results of comparative fracture endurance tests of nitro carburized pairs with unground teeth having the basic rack profile KS and of involute pairs.

Pairs with the basic rack profile KS			Involute pairs		
T_2^E, $N \cdot m$	$N_{F1} \cdot 10^{-6}$ before fracture	State of surfaces	T_2^E, $N \cdot m$	$N_{F1} \cdot 10^{-6}$ before fracture	State of surfaces
3555	10.45	Clean	2539	4.80	Weak pitting
	11.91	- " -		60.70	Considerable pitting
	51.00	- " -	2641	1.15	Clean
4063	2.27	- " -	3047	9.20	Lineage pitting
	4.36	- " -		54.00	Vast multiple pitting

The series of comparative fracture endurance tests of NG with the basic rack profile KS and involute gearing with carburized teeth made of steel 12HN2A under accelerated loads (corresponding to the inclined segment of the fatigue curve), carried out at VNIIReduktor (Kiev), showed that the extreme torque, reduced to $4 \cdot 10^6$ stress cycles, is $T_{F\,lim\,2}^E = 3295$ $N \cdot m$ for NG and 2498 $N \cdot m$ for involute gearing, makes the difference of 1.32 times. Taking into account the stated considerable scatter of testing results, the difference according to allowable torques is significantly greater.

In order to check the possibility of the basic rack profile KS effective application for thermally improved gearing, a number of tests for gearing with this basic rack profile and the degree of accuracy 9-9-8 [40], made of steel 40HN-U [129], was carried out. The obtained results indicate that the load-bearing capacity of thermally improved gearing with the basic rack profile KS is close to the load-bearing capacity for analogous gearing with the basic rack profile RGU-5 and, therefore, it exceeds the load-bearing capacity of gearing with the basic rack profile according to the standard GOST 15023-76 (see below) considerably.

In accordance with the noted above decision of the Coordinating Board at VNIIReduktor, a series of tests for nitro carburized pairs with unground teeth with basic rack profiles DLZ-0.7-0.15 and DLZ-1.0-0.15 [129] was carried out. The experiment results were very similar to the described above for the gearing with the basic rack profile according to the standard GOST 15023-76: a considerable scatter of results, their weak predictability and incapability to perform the correlation analysis [177]. Along with numerous tooth edge fractures, progressive pitting of their active flanks in the form of large cavities was observed.

Analysis of results of long-term tests of NG with surface hardened teeth with various basic rack profiles and comparison with the results shown by involute analogs allowed the Coordinating Board, as it was mentioned above, to standardize the basic rack profile KS for NG with high-hardened flanks in the branch [177], and then within the CIS (Commonwealth of Independent States) as the Interstate Standard GOST 30224-96 [39].

Results of the investigation, performed at "OOO "Reduktor", are of considerable interest for thermally improved Novikov pairs with the basic rack profile RGU-5 and tooth hardness $H_{HB} < HB\ 350$, which are not finally machined after gear milling.

Four tests were carried out in casings of double-stage gearboxes Ts2U-250 for gear pairs with $m = 3.15\ mm$, $z_1 = 33$, $z_2 = 62$, $b_w = 40\ mm$, $\beta = 20.747^\circ$ (high-speed stage) and with

m=5 mm, $z_1 = 18$, $z_2 = 75$, $b_w = 63$ mm, $\beta = 21.565°$ (low-speed stage), having the tooth hardness H_{HB1} = HB 302, H_{HB2} = HB 269, and also 6 tests for gear pairs with m=6.3 mm, $z_1 = 13$, $z_2 = 62$, $x_1^* = -0.093$, $x_2^* = -0.188$, $b_w = 63$ mm, $\beta = 20.223°$, H_{HB1} = HB 270, H_{HB2} = HB 230 (low-speed stage).

Letus describe briefly the purpose of these tests. The produced gearboxes Ts2U-250 and Ts3U-250 comprising NG with the basic rack profile taken according to the standard GOST 15023-76, with the tooth hardness $H_{HB} <$ HB 320, stopped satisfying the increased requirements given to gearboxes of this dimension type with respect to the specific material capacity (ratio of the gearbox mass to the taken torque). Then the decision was made to provide these gearboxes with nitro carburized involute gearing. However, it was established soon, that it was rather difficult to provide the qualitative nitro carburizing for the large-batch production of gears of the low-speed stage, because of a big overall dimension and mass. In this case, it was decided to try to mount the NG with the basic rack profile RGU-5 and the tooth hardness of the gear $H_{HB} \leq$ HB 230 after martempering at the low-speed stage of the gearbox.

The tests were performed in two stages. At the first stage, the load-bearing capacity was compared of pairs with m=3.15 mm and m=5 mm with basic rack profiles RGU-5 and according to the standard GOST 15023-76, of the same accuracy and tooth hardness (H_{HB1} = HB 302, H_{HB2} = HB 269 – see above). Long-term tests of pairs with the basic rack profile RGU-5 showed that $T^E_{HP2} = 1500\ N \cdot m$ ($N_{H1} = 98.6 \cdot 10^6$) for the high-speed stage and $T^E_{HP2} = 6250\ N \cdot m$ ($N_{H1} = 52.5 \cdot 10^6$) for the low-speed stage, after that they were accepted serviceable for further operation according to the tooth flanks state, though there was a feebly marked pitting. Since there were no tooth fractures, the noted torques are allowable according to fracture endurance of the gearing. Teeth of analogous pairs with the basic rack profile according to the standard GOST 15023-76 and with m = 5 mm were broken under the loads, 1.3...1.4 times less than the noted ones. After data analysis, including the results of previous numerous tests of gearing with the basic rack profile according to the standard GOST 15023-76, it was established that the load-bearing capacity of gearing with the basic rack profile RGU-5 is at least 25% more, than of similar gearing with the basic rack profile according to the standard GOST 15023-76.

Further, due to the increase of the gearing module at the low-speed stage up to 6.3mm (keeping the same overall dimensions of the gearing), it became possible to reduce the gear tooth hardness down to the desired level of HB 230, without reducing the load-bearing capacity, owing to it, the manufacturing process of gear production was considerably reduced in cost.

Long-term tests of the low-speed pair with m = 6.3mm with the basic rack profile RGU-5 in gearboxes 1Ts2U-250 and 1Ts3U-250, the high-speed stage of which was provided with nitro carburized involute pairs, showed its high load-bearing capacity according to the contact endurance ($T^E_{HP2} = 5008\ N \cdot m$, $N_{H1} = 80.5 \cdot 10^6$). No cases of tooth fracture were observed here and flanks were in satisfactory state (Fig. 21*a*). Tests were continued under the load T^E_2

$= 10000\ N \cdot m$ and they also did not lead to fracture, however, after $5.5 \cdot 10^6$ cycles, a visible pitting of pinion flanks appeared (Fig. 21*b*). Then the pinion was made of steel 25HGM, nitro carburized and hardened up to $H_{HRC1} = HRC\ 56$ and tests were repeated under the load $T_2^E = 10000\ N \cdot m$. The absence of fracture in all tests allows to take this load as the extreme one according to the fracture endurance of the gear.

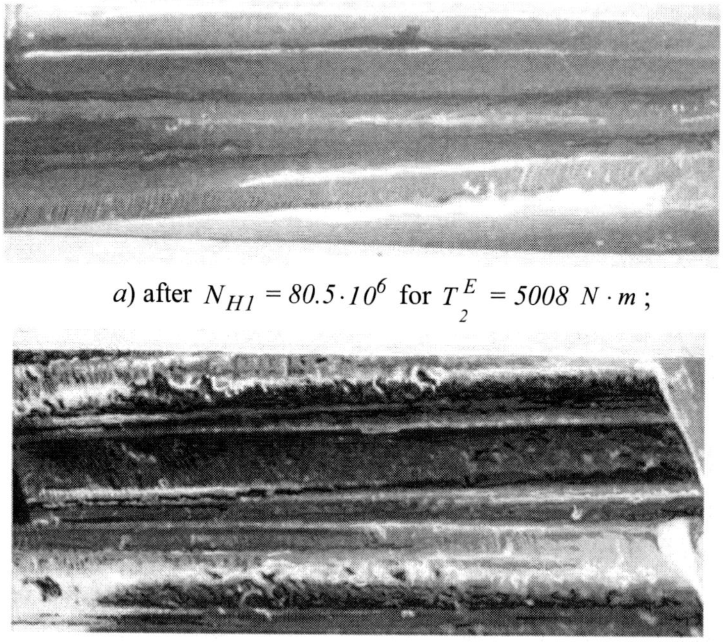

a) after $N_{H1} = 80.5 \cdot 10^6$ for $T_2^E = 5008\ N \cdot m$;

b) after $N_{H1} = 5.5 \cdot 10^6$ for $T_2^E = 10000\ N \cdot m$

Figure 21. State of tooth flanks of the thermally improved pinion $z_1 = 13$.

After typical and evaluation tests of the pair *13/62* with $m = 6.3mm$ of the nitro carburized pinion ($H_{HRC1} = HRC\ 56$) and thermally improved gear ($H_{HB2} = HB\ 230$) with basic rack profile RGU-5 was mounted at the low-speed stage of double-stage 1Ts2U-250 and three-stage 1Ts3U-250 gearboxes, produced at "OOO "Reduktor" in large batch, providing the increase of their load-bearing capacity and reduction of the specific material capacity *1.25* times compared with Novikov pairs, made on the basis of the basic rack profile according to the standard GOST 15023-76. Hob production for tooth-cutting of Novikov gears with basic rack profiles RGU-5 and according to the standard GOST 30224-96, was mastered at toolmakers plants MIZ (Moscow) and SIZ (Ekaterinburg).

Relying on the described above, one can conclude that the most efficient area of NG application is high-stressed power drives, and it is advantageous here to use the maximum possible gearing module, since not only bending, but also contact endurance of teeth is increased in this case, and the gearing sensitivity to radial errors is also decreased.

This conclusion, as we will see further (see Chapters 11, 12, 13, 16), is shown in techniques of strength analysis and is applied in recommendations for NG design.

Chapter 9

INDUSTRIAL APPLICATION AND RESULTS OF CYLINDRICAL NOVIKOV GEARING OPERATION

Industrial development of NG OLA began in 1956, and TLA gearing – in about 1960 [173]. More than 300 scientific-research organizations and plants joined this activity, however, along with the achieved success of NG implementation, the negative result was obtained in certain cases, which is explained by the deficiency of some original statements and the desire to obtain the maximum increase of the load-bearing capacity quickly without performing necessary experimental and theoretical investigations [82].

The first sample gearboxes with NG OLA were manufactured at Parkhomenko Lugansk Plant [173], where the batch production of gearboxes for the coal-mining industry had been mastered since 1960. When substituting involute gearing with NG OLA, the gearbox mass was reduced by $(15...20)\%$.

Since 1957, the development of NG OLA began at the Rostov Machine-building Plant [60]. Several versions of basic rack profiles were proposed and gearing, produced on the basis of these profiles, was tested. In 1958, the gearbox with NG was put into batch production. There appeared the possibility [34] to decrease the hardness of gears from $H_{HRC} = HRC\ 56$ (involute version) down to $H_{HB} = HB\,180$ (version of NG).

One of the first applications of NG is the batch production of gearboxes of coal-mining conveyors [54] at the Kharkov Machine-building Plant “Svet shakhtera”. Initially, the basic rack profile according to MN 4229-63 (OLA), and then according to the standard GOST 15023-76 (TLA) was used, and the long-term experience of operation showed that the thermally improved NG successfully replaced the hardened involute gearing.

After obtaining the results of theoretical investigation [224], testing and implementation [137, 140] of TLA gearing, a new stage of NG application began. Practice showed that, in overwhelming majority of cases,the TLA gearing surpasses the OLA gearing according to the load-bearing capacity. Thus, in gearboxes of underground conveyors, the substitution of OLA for TLA gearing enabled the increase of the gearbox capacity by more than 30% [54], and substitution of OLA for TLA with the basic rack profile according to the standard GOST 15023-76 in the drive of coal-cleaning and mining machines [56] allowed the reduction of the gearbox mass by $(10...12)\%$.

NG of low hardness on the basis of the basic rack profile Ural-1K (one of the prototypes of the basic rack profile according to the standard GOST 15023-76) is successfully applied in drives of underground drag-type conveyors and pick-and-place devices, produced at the Skopinsk Machine Plant (Ryazan region).

Nowadays, a considerable beneficial experience is accumulated on the industrial application of NG TLA with flank hardness under *HB 320* in many machines, including gearboxes of coal-mining lift machines [184] and especially in drives of oil pumping units. Their production was first organized in Azerbaijan [80] and then was successfully continued in Russia (at Izhevsk "OAO "Izhneftemash" and Izhevsk "OOO "Reduktor") with the new basic rack profile according to the standard GOST 30224-96 [39], also showing good results for teeth with low hardness. As a rule, implementation of NG leads to the rise of load-bearing capacity and durability of the power drive or reducing the gearbox mass. Fig. 22 shows the driving gearbox of the oil pumping unit, provided with thermally improved herring-bone NG.

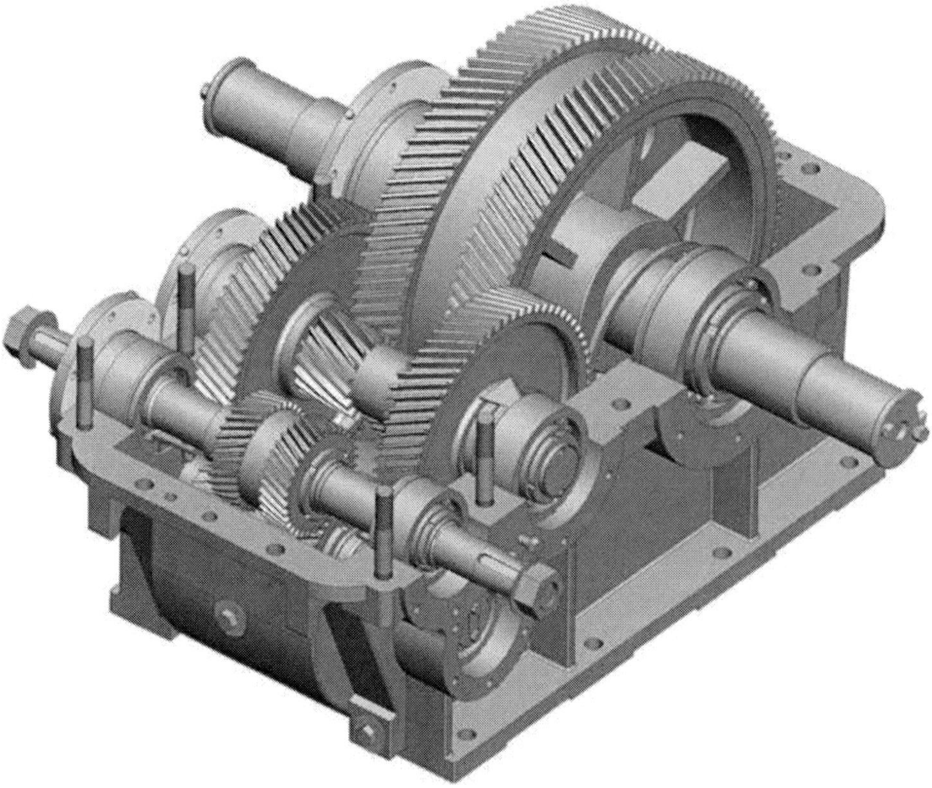

Figure 22. Driving gearbox of the oil pumping unit with thermally improved herring-bone Novikov gearing.

The example of operation of two large-size herring-bone NG with the basic rack profile RGU-5 is also demonstrative, they were designed in 1977-1978 at the Southern Federal University (SFU) and produced at the Syzran Turbine Construction Plant. These gears were intended for gearboxes of the drive of coal-pulverizing mills of Sh-25A type, applied at the Thermal Power Plant-22 Mosenergo. It was decided to mount the gearing instead of the involute one, the width of the half of herring-bone was reduced *1.5* times, which gave the economy of working hours of gears production and mass reduction of one pair by *0.7 ton*. The main parameters of the pair were: $m=8\ mm$, $z_1=33$, $z_2=146$, $a_w=800\ mm$, $b_w=130\ mm$ of each half of herring-bone (instead of *200mm* for involute gearing), $\beta = 26^\circ\ 29'30''$. The pinion was made of steel 40HN and thermally improved up to the hardness $H_{HB} = HB\,250$, the gear tread was made of steel 45 in as-supply condition ($H_{HB} = HB\,215$). The tangential speed of gearing rotation was $v = 11.5$ *m/s*. Both pairs were operating for about 10 years, having worked during approximately $3 \cdot 10^9$ cycles at the pinion, which was *2.5...3* times more than the average operating time of the replaced involute analogs, and they were demounted as a result of the natural tooth wear. Annual examinations did not reveal any visible damages of tooth flanks.

NG became widely spread in cylindrical gearboxes of general engineering purposes. These gearboxes are batch produced with gears, made on the basis of the basic rack profile according to the standard GOST 15023-76 at the Maykop Gearbox Plant and Izhevsk "OOO "Reduktor".

The latter enterprise took an active part in the investigations of the load-bearing capacity of NG [81]. Even in the beginning of 1960s, it was established that the load-bearing capacity of gearboxes of RTsD type, equipped with NG with flank hardness under *HB 300,* is approximately *1.5* times greater than that of the same gearboxes with involute gearing, operating continuously.

However, it is also important to establish the influence of variable loads on the working capacity of NG, since about *30%* of cylindrical gearboxes operate under variable loads. The information insufficiency about the influence of the loading mode limited the implementation of the new gearing in general engineering purpose gearboxes.

In this connection at Izhevsk "OOO "Reduktor", investigations were carried out [20, 21, 23] of gearboxes with TLA gearing under variable operation mode. Their high working capacity was then proved under variable loads and short-term overloads and it was established that, in these cases, they transmitted loads *1.3...1.4* times greater than under continuous modes of operation.

On the basis of the investigation results, gearboxes RTsD-250, RTsD-350 and RTsD-400 with NG were developed, and their batch production started in 1968.

In recent years, gearboxes of Ts2U series were developed instead of gearboxes of RTsD series at Izhevsk "OOO "Reduktor". Since 1985, and until today, gearboxes Ts2U-250, 1Ts2U-250 and 1Ts3U-250 with NG are being produced, which are applied mainly in hoisting cranes, working under variable operation modes. More than a half of million gearboxes Ts2U-250, 1Ts2U-250 and 1Ts3U-250 has been manufactured since 1985, and, as it was noted in Chapter 8, application of the basic rack profile RGU-5, instead of the basic rack profile according to the standard GOST 15023-76 at the low-speed stage of gearboxes 1Ts2-U-250 and 1Ts3U-250, led to the increaseof the load-bearing capacity of these gearboxes by *25%* compared with the replaced gearboxes Ts2U-250 and Ts3U-250. Gear-motors MTs2S-100N with NG RGU-5 are also successfully manufactured.

If the tangential speed in the gearing is more than *90 m/s*, gearing with the basic rack profile YuTZ-65 is successfully applied in the drive of centrifugal compressors. As the result of the total experience of testing, operation and performed investigations [222], a special design office for compressor engineering and the Kazan Compressor Plant developed a plant standard "Cylindrical speed-up gearboxes with Novikov gearing", which was comprised of 5 dimension types of speed-up gearboxes. Their main parameters were: the transmitted power from *375* to *7550 kW;* gear ratio *1.25...4;* center distance *200...425 mm;* axial overlap ratio *2...5;* rotational frequency of the shaft connected with the drive $3000\ min^{-1}$.

Application of NG in standard speed-up gearboxes allowed for the decrease of the mass and overall dimensions of the latter considerably, since the load-bearing capacity of high-speed NG with flanks hardness under *HB 320* is *2...3* times greater than the load-bearing capacity of involute gearing [222]. Many speed-up gearboxes with NG TLA, manufactured at the Kazan Compressor Plant, have the non-failure operating time of *87-100 thousands of hours,* that is, *10...12* years without replacing the gear pair.

Similar conclusions concerning the relation between load-bearing capacities of Novikov and involute gearing were made at the Chernomorsk Ship-building Plant [10], where gearing with the basic rack profile YuTZ-65 with tooth hardness $H_{HB} = HB\ (280...320)$ had been applied since 1966 in high-speed gearboxes of turbine-compressor stations of K-250-61-1 type. As the result of the implementation of this gearing, the manufacturing process of gear production has been simplified and reduced in cost due to the elimination of a number of mechanical and heat treatment operations; vibro-acoustic characteristics have been improved; the gearing lifetime has been increased *2...3* times as the previously applied involute gears often failed because of tooth flanks pitting.

Advantages of NG can be implemented, in some cases, only with the proper quality of gear production. Thus, under considerable manufacture and assembly errors, the pitting of tooth flanks of NG was observed in compressor machines with an operating time of *7348 hours* [142]. In case of the proper gearing production quality and the absence of operation mode violation, their satisfactory operation has been registered during *37000 hours* [25]. The experience of the Khabarovsk Plant of Power Machine-building, producing the speed-up gearboxes of centrifugal compressor machines in batch, proves the advantages of NG compared with the involute one when applying them in the high-speed drive.

The company “SPIN” (Orel) has been recently organized; it produces large gearboxes with high-hardened NG for metallurgical equipment.

All of the described above is referred to the experience of application of gearing with tooth surface hardness up to *HB 320* in Russia. Application of NG with high-hardened tooth flanks is still spread to a smaller extent, which is mainly related to insufficiently mastered technique of high-quality manufacture and control of this gearing.

In 1969, at the Uritsky Trolleybus Plant (Engels), cylindrical NG with the basic rack profile Don-63 was implemented into batch production; it was used in gearboxes of the rear axle of the trolleybus ZIU-5 [58, 196]. The gearing was manufactured according to the $8\text{-}9^{th}$ degree of accuracy with hardness of carburized teeth $H_{HRC} = HRC\ (58...64)$. The increase of durability of *8* times compared with involute gearing with ground teeth of the 7th degree of accuracy had been achieved.

The central gearbox VR-126M with NG is used in the Russian helicopter K-126.

Gearboxes Ts2U-160 of the pilot batch are successfully operated, both stages of them are provided with nitro carburized gear pairs with the basic rack profile RGU-5, which is intended for high-hardened gearing of general engineering application. (Tests of gearing with this basic rack profile are in details described above in Chapter 8). Izhevsk “OOO “Reduktor”, together with SFU, developed the unified series of progressive gearboxes with nitro carburized NG on the basis of the basic rack profile according to the standard GOST

30224-96, hob production for cutting gear teeth have been successfully mastered at the Moscow Toolmakers Plant.

The first industrial operation of high-hardened NG with the basic rack profile according to the standard GOST 30224-96 was implemented in drives of French blowers DNH-110 at the Nikolaevsk Lumnite Plant as import substitution of expensive proprietary involute pairs, where very encouraging results were obtained.

Enough attention is also paid to NG abroad [79, 83, 145, 156, 253, 255, 256, 260, 263, 266]. Investigation works are carried out in the USA, Japan, China, India, Germany, Belgium and other countries, in a number of cases showing the considerably increased load-bearing capacity of NG (compared with involute analogs), the increased thickness of the oil film in contact, and also positive results for tangential speeds up to 120 *m/s*.

Thus, for example, [156] gives information about coming into the market and practical application of gear pairs with NG. In the opinion of the authors of the publication, a wide application of this gearing type is limited by the absence of well-elaborated manuals on its design and corresponding standards, and also by strict requirements to the accuracy of the center distance. The increase is registered of the load-bearing capacity of NG with respect to the contact strength several times compared with the involute gearing having gear made of thermally improved steels.

In [165], the results are given of application in Chinese People's Republic of gear pairs with NG TLA with high-hardened teeth at motor ships with the main power plant having the capacity *1950 kW*. Its main characteristics were: $N=970\ kW$, $a_w = 560\ mm$, $m=7\ mm$, $z_1 = 39$, $z_2 = 117$, $\beta = 12^{\circ}50'17''$, $b_w = 260\ mm$, $n_1 = 428\ min^{-1}$, quenching, nitriding, $H_{HRC} = HRC\ 60$. Parameters of the basic rack profile were: $\alpha_k = 30^{\circ}$, $h_a^* = 0.6$, $h_f^* = 0.8$, $\rho_a^* = 0.75$, $\rho_f^* = 0.9$. The accuracy of the profile was 7th, the degree of standard was IB 179-60. NG allowed to improve vibration characteristics of the drive and increase the efficiency. The conclusion was made that operation experience of such gearing made it possible to predict its application in drives of even greater capacity.

In the USA (the company "DARCO") and in China, gearboxes with NG for oil pumps, exported to a number of countries, have been successfully applied for many years.

There is certain information on performing works on the possibility to apply the Novikov power train in gearboxes of helicopters, produced by the English company "WESTLAND HELICOPTER".

Therefore, a rather considerable experience has been accumulated nowadays on the application of NG with the tooth hardness under *HB320*. Positive results have also been obtained, indicating the progressive character of application of this gearing in pairs with high-hardened teeth in heavy-loaded drives of various machines [106].

Chapter 10

Design Loads in Cylindrical Novikov Gearing

For Novikov gearing, similarly to the involute one [43], the following types of checking strength analyses are used, necessary to evaluate their working capacity (Table 12).

Table 12. Types of Novikov gearing analysis.

Type of analysis	Purpose
Tooth contact strength analysis	
Contact endurance of active flanks	Prevention of fatigue pitting of active tooth flanks
Contact strength of active flanks under the action of maximum load	Prevention of residual deformations or brittle fracture of the surface layer
Depth contact strength	Prevention of the depth contact fracture in the body of teeth with surface hardening
Tooth bending strength analysis	
Bending endurance	Prevention of tooth fatigue fracture
Bending strength under maximum load	Prevention of residual deformations or brittle fracture of teeth

The dependence between forces in the gearing and the torque T, necessary for all calculations, is expressed by the following relationship:

tangential force $F_t = 2000T\cos\beta/(mz)$; (10.1)

normal force $F_n = 2000T/(mz\cos\alpha_k)$; (10.2)

radial force $F_{rad} = 2000T\tan\alpha_k/(mz)$; (10.3)

axial force $F_x = 2000T\sin\beta/(mz)$. (10.4)

In order to determine design forces, corresponding to one contact area, the given formulas should contain the design torque, related to the input torque T by using load coefficients [123]:

$$T_H = K_H T, \tag{10.5}$$

$$T_F = K_F T, \tag{10.6}$$

where coefficients K_H (used in contact strength analysis) and K_F (used in bending strength analysis) are as follows:

$$K_H = K_A K_{H\sigma} K_{Hv} K_\lambda K_f; \tag{10.7}$$

$$K_F = K_A K_{F\sigma} K_{Fv}. \tag{10.8}$$

Constituent multipliers of (10.7), (10.8) are:

K_A - coefficient of the external dynamic load, which can be assigned the same as in involute gearing calculations [43];

$K_{H\sigma}, K_{F\sigma}$ - coefficients that take into account the non-uniform distribution of contact and bending stresses correspondingly among contact areas;

$K_{Hv,} K_{Fv}$ - coefficients that take into account the dynamics of gearing when calculating the contact and bending endurance, correspondingly;

K_λ - coefficient that takes into account the adaptability of the gearing;

K_f - coefficient of accounting the friction forces in contact.

Prior to giving the detailed explanation of all these coefficients - components of (10.7), (10.8), letus consider the basic problem – the character of load distribution among contact areas.

10.1. Non-Uniform Load Distribution among Contact Areas

The load-bearing capacity of a gearing depends essentially on the non-uniform distribution of loads and stresses among contact areas.

A number of works [65, 90, 91, 196] is devoted to problems of load distribution among contact areas, however, explicit design relations, taking into account the gearing operation in real-life environment, have not been obtained there. That is why it is very urgent to study this problem for NG, since it enables the considerable strength analysis improvement and specification of the rational application field for this type of gearing, especially with high-hardened active flank teeth.

Load distribution among contact areas can be characterized by a certain coefficient K_{Ti}, representing the ratio of a certain partial load T_i, acting on the "i"-th contact area, to the total input load T, that is,

$$K_{Ti} = T_i / T. \tag{10.9}$$

As it is known, two coefficients are used in practice of involute gearing calculation: $K_{H\alpha}$ ($K_{F\alpha}$), taking into account the load distribution between teeth, and $K_{H\beta}$ ($K_{F\beta}$), taking into account its non-uniform distribution along the length of contact lines, these coefficients are introduced as the multipliers in design formulas [43]. Application of the noted coefficients in the practice of NG calculation during many years [140, 141, 176 and oth.] is wrong, since the specific feature of NG is not considered here at all.

Features of theoretical point NG, having one or several contact areas under the load, located at different teeth and shifted with respect to each other along the gearing axes, do not allow to present the coefficient K_T as the product $K_\alpha \cdot K_\beta$. In order to characterize the non-uniform load distribution among contact areas, it is reasonable here to introduce a certain unified coefficient $K_T = K_{\alpha\beta}$, equal to the ratio of the load, corresponding to the considered contact area and a certain fixed constant load, which is convenient to be presented as the summarized input load T [100].

The coefficient K_T depends on many factors, the main ones being : the total elastic compliance W_δ of teeth, tooth elastic displacement W_k, caused by the compliance of mating construction parts (shafts, bearing supports and so on), errors Δ of gears manufacture and assembly, tooth flanks run-in.

In order to determine the coefficient K_T, the design model [65] is applied, described by the system, containing equations of the elastic equilibrium and consistency of contacting teeth elastic displacements:

$$\begin{cases} W_{\delta j} + W_{Kj} + \Delta_j = W_{\delta(j+1)} + W_{k(j+1)} + \Delta_{(j+1)}; \\ \sum_{i=1}^{n} T_i = T, \quad j = 1,2,..n-1. \end{cases} \tag{10.10}$$

The summarized contact and bending-shear displacement W_δ of the tooth in the contact point is determined according to the approximating dependence

$$W_\delta = K_{uk}(K_b T_2)^{0.642} m^{-0.926} (\rho_\beta^*)^{-0.299} z_2^{-0.67} (\cos\alpha_k)^{-1.642} (\cos\beta)^{-1}, \tag{10.11}$$

obtained as the result of simulation of the task solution for three-dimensional stress-strain tooth state [88, 89].

When considering the elastic displacement W_k of design parts, the dependence [185] is used:

$$W_{ki} = [a_{ij}^*]\{F_{nj}\} + \Delta_{pi}, \tag{10.12}$$

where $[a^{*}_{ij}]$ is the matrix of influence coefficients of the shaft elastic compliance, Δ_{pi} is the displacement in the "*i*"-th contact point as the result of bearings deformation under the force action.

Kinematic (without load) initial clearance in contact points, caused by gears manufacture and assembly errors and defined in the stohastic aspect, is stipulated by the obliquity f_{xr} and misalignment f_{yr} of axes, errors $F_{\beta r}$, f_{fr} of the tooth line direction and its profile, and by the deviation f_{ptr} of the tangential pitch [100].

In (10.11), K_{uk} is the coefficient, depending on the basic rack profile type: for the basic rack profile according to the standard GOST 15023-76 $K_{uk}=0.0692$, for the basic rack profile RGU-5 $K_{uk}=0.0651$, for the basic rack profile taken according to the standard GOST 30224-96 $K_{uk}=0.0641$.

The coefficient K_b takes into account the influence of the face surface (that is, the vicinity of the contact area to the face surface). As it has been established theoretically [24], and proved experimentally [65], this coefficient is found within the range from *1* to *2*, varying exponentially.

Coefficients K_{Ti} are obtained in the form of (10.9) as the result of the system (10.10) solution.

10.2. Accounting the Non-Uniform Distribution of Bending Stresses among Contact Areas

Taking into account that the coefficient K_b of the face surface influence is similar when determining the displacement (10.11) and bending stresses [65, 88], one can write:

$$K^{0}_{F\sigma} = K_T K_b, \tag{10.13}$$

where $K^{0}_{F\sigma}$ is the initial (before run-in) coefficient of bending stresses concentration.

The numerical simulation in a wide range of geometrical parameters and loads allowed to establish that the coefficient $K^{0}_{F\sigma}$, which reached maximum values at face surfaces practically in all cases (which is proved by face fractures of teeth), can be approximated by the following rough relation:

$$K^{0}_{F\sigma} = A\lambda^{a}\psi^{b}_{bd}u^{c}m^{d}, \tag{10.14}$$

where $\psi_{bd} = b_w / d_1$.

The parameter λ is determined according to the accuracy/rigidity criterion:

$$\lambda = \gamma_c\sqrt{F^{2}_{\beta 1} + f^{2}_{pt1}} / W_{\delta}, \tag{10.15}$$

where γ_c is the coefficient (see Table 9), depending on the accepted "risk level" under the assumption of the normal law of errors; $F_{\beta 1}, f_{pt1}$ are the pinion tolerances for the error of the tooth line direction and deviation of the tangential pitch for the given degree of accuracy correspondingly [40].

The constituent value W_δ of (10.15) is calculated according to (10.11) for $K_b = 1$.

By analogy with [43], but with account of the NG specific character, the coefficient $K_{F\sigma}$ of bending stresses concentration after run-in can be expressed by the following structural formula:

$$K_{F\sigma} = K_{Fc} + (K^0_{F\sigma} - K_{Fc})K_w, \tag{10.16}$$

where K_{Fc} is the coefficient of bending stresses distribution among contact areas for the "ideally precise" gearing for the absolutely rigid design, K_w is the coefficient, considering the teeth run-in [43] and depending on the teeth tangential velocity and tooth flanks hardness.

Coefficients A, a, b, c, d and the parameter K_{Fc} are taken according to Table 13. Calculations show that the influence of shafts rigidity can be neglected for stages of the gearboxes with symmetrical or close to symmetrical arrangement of gears with respect to bearing supports, and in this case, parameters of Table 13 are used for $n_k = 0$ (that is, $b=0$, $c=0$, $d=0$). Inother cases, when the shaft rigidity must be taken into account, parameters of Table 13 are used for $n_k = 1$. Table 13 also gives the values $(K_{F\sigma})_{min}$, and the design values $K_{F\sigma}$ can not be taken under these values. Deviations Δ_F, % of the design values according to (10.14) from the precisely calculated ones by means of the specially developed software "LSZPVK" are also given in Table 13. When using Table 13, it is necessary to pay attention to comments given below the Table.

The parameter α is used in those cases when the numbers of degrees of accuracy of the gearing according to the smoothness (k_p) and contact (k_k) ratings do not coincide, that is, $k_p \neq k_k$. In this case, the parameter λ is calculated according to the relation:

$$\lambda = \alpha\lambda_1 + (1-\alpha)\lambda_2, \tag{10.17}$$

where λ_1 is calculated for the same values k_{p1}, k_{k1} of the relative gear, equal to the degree of accuracy k_p of the given gear, and λ_2 is calculated for the same values k_{p2}, k_{k2} of the relative gear, equal to the degree of accuracy k_k of the given gear.

When calculating $K^0_{F\sigma}$ for a herring-bone gearing, parameters ψ_{bd}, $F_{\beta 1}$ and ε_β are taken as for the helical gearing with the operating width of toothing b_w, equal to the twice toothing width of the half of the herring-bone.

Table 13. To calculate the coefficients of bending and contact stress concentration.

ε_β	K_{Fc}	K_{Hc}	$K_{F\sigma\,min} = (K_{H\sigma c})_{min}$	$n_k=0$				$n_k=1$						
				A	a	α	$\pm\Delta_F$,%	A	a	b	c	d	α	$\pm\Delta_F$,%
0.9	0.77	0.67	1.0	1.240	0.368	0.75	1.5	1.327	0.342	-	-	-	0.75	2.7
1.0	0.55	0.45	0.5	1.089	0.469	0.63	2.0	1.161	0.436	-	-	-	0.63	3.3
1.1	0.50	0.40	0.5	1.034	0.501	0.62	1.9	1.118	0.466	-	-	-	0.62	3.6
1.2	0.48	0.39	0.5	1.000	0.516	0.62	1.8	1.100	0.480	-	-	-	0.62	3.4
$1+\varepsilon_q$	0.34	0.28	0.333	0.847	0.609	0.55	3.3	1.427	0.436	0.302	-0.176	-0.020	0.55	7.6
2.0 *)	0.29	0.24	0.25	0.786	0.641	0.52	3.0	1.569	0.413	0.528	-0.296	-0.027	0.52	9.0
2.1	0.28	0.22	0.25	0.766	0.652	0.51	3.2	-	-	-	-	-	-	-
2.2	0.27	0.22	0.25	0.752	0.657	0.51	3.3	-	-	-	-	-	-	-
$2+\varepsilon_q$	0.22	0.18	0.2	0.683	0.691	0.48	4.1	1.397	0.346	0.554	-0.299	-0.032	0.49	11.3
3.0	0.20	0.15	0.167	0.667	0.700	0.48	4.0	1.113	0.351	0.695	-0.248	-0.037	0.50	9.5
$3+\varepsilon_q$	0.19	0.15	0.143	0.627	0.711	0.47	5.6	1.111	0.366	0.697	-0.363	-0.047	0.51	10.2
4.0	0.19	0.14	0.125	0.582	0.733	0.48	7.1	0.935	0.380	0.816	-0.368	-0.060	0.52	9.2

Notes: 1. Reference design values of ε_β are designated by the bold type; the values $K^0_{F\sigma}$ calculated for them according to (10.14) are also taken for the next after reference values of ε_β up to the nearest ε_β values; 2. *) is a reference one only for $n_k=1$; 3.In ranges $\varepsilon_\beta=1.0\ldots1.2$ and $2.0\ldots2.2$ a linear interpolation of table parameters is allowable.

10.3. Accounting the Non-Uniform Distribution of Contact Stresses among Contact Areas

It is considered conventionally that when designing NG, the main attention should be paid to the tooth fracture strength, since this type of gearing is organically featured by the high contact strength of contacting surfaces. Such an approach became, in due time, the reason of appearance of rather low basic rack profiles, providing the increased fracture strength of teeth (for rather precise manufacture of the gearing). However, in certain cases, it is impossible to decrease the contact strength [129]. Insufficient attention to the tooth flanks contact strength can negatively affect the load-bearing capacity of NG with any hardness of tooth flanks [129]. That is why the ability of the designer to estimate very objectively the contact strength of NG teeth will allow to use reserves of this progressive gearing in a wider sense, by assigning optimal parameters.

Fig. 23 shows the displacement of the contact pattern *1* along the arrow towards the tooth end in the process of engagement, causing in this position the "peak" of contact stresses, or the so-called "boundary effects".

In the well-known investigations, devoted to the problem of "boundary effects" under the contact interaction of bodies (for example, [244]), the results are given for the quarter space, that is, the special case, when the dihedral angle at the edge is equal to $\pi/2$, the decision is made only up to displacements, the design error of which increases as approaching the edge due to the approximate character of the obtained Green function. However, for more thorough investigation of the "boundary effects" task, it is necessary to know not only displacements, but also the SSS in the contact near the edge (tooth edge) with the acceptable accuracy, under the condition, that the angle of surfaces intersection near the edge differs significantly from $\pi/2$.

According to the described above, as an illustration of the model of tooth flanks contact interaction of NG, the task was stated and solved on indentation of the rigid die having the shape of the elliptical paraboloid with the main radii of curvature R_{min}, R_{max} into the elastic spatial wedge with the auxiliary dihedral angle α at the edge [119]. Contact displacements, normal pressures and effective (octahedral) stresses in the contact were determined here by the stress tensor.

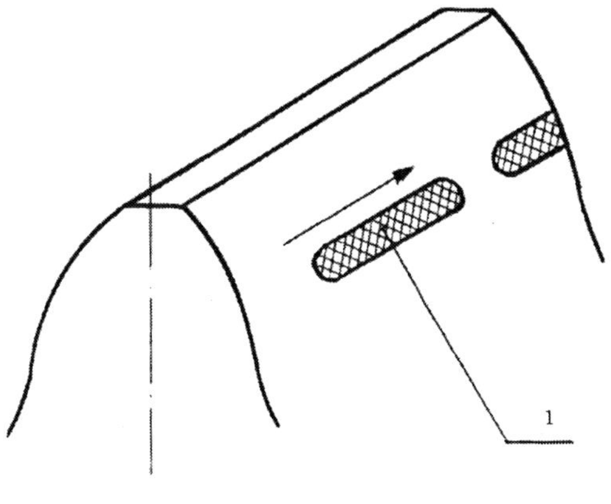

Figure 23. Scheme of contact pattern displacement towards the face edge in engagement.

Not giving the mathematical conception here, note that the described contact task has been solved with application of the method of Hammerstein-type boundary non-linear equations [29], the kernel of the integral equation is regularized both outside the edge and at the edge of the wedge. It was established that "risky" stresses appear from that side of the tooth surface, which generates the obtuse angle $\alpha = \pi / 2 + \beta$ with the face surface of the gear rim.

Table 14 shows the calculation results of coefficients $K_{\sigma m}$, representing the ratio of face effective contact stresses to such stresses for the half-space, depending on the angle α and the relation $C_{\alpha\beta} = R_{min} / R_{max} = \rho_\alpha / \rho_\beta$.

Table 14. Values of $K_{\sigma m}$

$\alpha,^\circ$	Values of $K_{\sigma m}$, for $C_{\alpha\beta}$							
	0.025	0.05	0.075	0.1	0.15	0.2	0.25	0.3
105	1.01	1.09	1.16	1.18	1.20	1.27	1.32	1.34
110	1.15	1.24	1.28	1.35	1.41	1.45	1.49	1.50
115	1.24	1.36	1.40	1.45	1.53	1.57	1.60	1.63

According to the results of [110], one can write for the face surface:

$$K_{H\sigma m} = 0.5 A_m K_{F\sigma}, \tag{10.18}$$

where, based on the data from Table 14 and the established power relation between the load and stresses (see Chapter 11 below):

$$A_m = 0.36 C_{\alpha\beta}^{0.16} \beta^{0.6}. \tag{10.19}$$

(The angle β is in degrees).

As investigations showed, maximum contact stresses can appear, unlike the bending ones, not only on the face surface, but also in the middle part of the tooth.

It was established by the numerical simulation, that "middle" initial coefficient $K_{H\sigma c}^0$ of contact stresses, equal to the corresponding coefficient K_T of load concentration, depends mainly on the accuracy of gearing and can be presented by the approximate relation:

$$K_{H\sigma c}^0 = 1.026 f_x^{-0.117} K_T = 0.513 f_x^{-0.117} K_{F\sigma}^0, \tag{10.20}$$

where f_x is the tolerance of the gearing axes obliquity for the assigned degree of accuracy [40].

Passing from $K^{0}_{H\sigma c}$ to $K_{H\sigma c}$ according to (10.16), that is, considering the flanks run-in, and taking the coefficient K_{Hc} according to Table 13, we will obtain finally

$$K_{H\sigma} = max\{K_{H\sigma c}, K_{H\sigma m}\}. \tag{10.21}$$

It should be kept in mind that the value $K_{H\sigma}$ must not be less than $(K_{H\sigma c})_{min}$, given in Table 13.

10.4. Accounting the Friction Forces in the Contact

It is known that during the operation of NG, which is out-pitch-point by definition, the sliding takes place between contacting flanks of the gear pair [158]. The influence of friction forces, appearing here on the SSS of flanks in contact is not reflected in the existing techniques of teeth contact strength analysis. It can be explained in addition by the absence of corresponding theoretical backgrounds. The available certain recommendations [96] of accounting the friction factor with reference on few experimental results had not been systemized.

Note that it is very difficult to consider the real pattern of friction in contact, since numerous factors affect it: roughness of interacting surfaces, inconstancy of friction factor and unknown law of its variation, presence or absence of lubrication, sliding velocity etc ., let alone the possibility of the partial sliding with cohesion, rolling and spinning [158]. In this connection, it is reasonable to obtain step-by-step decisions, gradually passing from simple to complex with improvement of mathematical methods, computer techniques and experimental data accumulation.

If we continue to apply the idea of the die indentation into the elastic space wedge as the model of NG flanks interaction [119], the task has been set to obtain theoretical results and develop methodical recommendations on gearing calculation on their basis with account of the action of not only normal, but also tangential loads (friction forces) in the contact area.

The task was solved under the following basic conditions [120]:

a) the friction force is produced by displacement of the die along the wedge surface in the direction, perpendicular to the edge of the latter;
b) the line of friction force action coincides with the line of action of the relative velocity vector of teeth sliding, that is, it is the intersection line of the plane, normal to axes of gears, with the plane, tangent to tooth surfaces in the contact point [158]; for the considered task the line of friction force action coincides with the direction of the minor axis of the contact ellipse;
c) the die, which is elliptical in the projection, is arranged with respect to the wedge in such a way that, as the die is moving, the major axis of the theoretical contact ellipse always remains parallel to the edge of the wedge and considerably exceeds the minor axis;
d) the velocity of the die motion is negligibly small compared with the velocity of elastic waves distribution;
e) there is a complete slippage and there are no adhesion areas between the die and the wedge;

f) the friction factor remains constant at the whole contact area, being the proportionality coefficient between normal and tangential forces (Coulomb friction);
g) there is no lubrication between interacting surfaces of the die and the wedge.

Under the enumerated conditions, the results of the task solution can be at the first stage put into the basis of the development of methodical recommendations on NG calculation refinement.

Not giving the detailed solution of the task (described in [30, 120]), note that the SSS was estimated by means of effective contact stresses, it was established here that with the increase of the friction factor f and the value $C_{\alpha\beta} = \rho_\alpha / \rho_\beta$, the values of effective contact stresses on the surface are also increased, according to one and the same law both near the face surface and far from it. This allows to consider the influence of friction, estimated by the coefficient K_f, independently on the gearing adaptability.

Investigations led to the following relation, introduced into the load-bearing capacity calculation of the gearing:

$$K_f = (1 + 60 f^2 C_{\alpha\beta}^{0.25})^{0.728}. \qquad (10.22)$$

10.5. Strength Aspect of the Gearing Adaptability

The concept of adaptability as the important characteristic of NG was given in Chapter 5.

It order to perform the strength analysis, it is very important to be able to estimate numerically the "peak" of contact stresses when the part of contact pattern runs beyond the boundaries of the tooth active height. There was no such estimation in the existing techniques, it was done for the first time in [110]. The solution was obtained within the model task on the indentation of the elliptical rigid die into the elastic spatial wedge [119], having the apex angle $\alpha = 135^\circ \ldots 140^\circ$, which corresponds, on average, to the angle of intersecting the tooth flanks with the surface of tooth tips. The qualitative relation was established here, shown in Table 15, between the part t of the semi-width b_H of the minor axis of the contact ellipse, exceeding the tooth boundaries, and the relation $K_{\sigma\lambda}$ of the originating effective contact stress to the same stress for the sufficient removal of the ellipse from the edge.

Table 15. Values of $K_{\sigma\lambda}$

t	Values of $K_{\sigma\lambda}$, for $C_{\alpha\beta}$							
	0.025	0.05	0.075	0.1	0.15	0.2	0.25	0.3
1	2.50	2.66	2.77	2.82	3.00	3.19	3.43	3.61
0.75	1.41	1.48	1.53	1.54	1.57	1.60	1.65	1.67
0.5	1.09	1.11	1.13	1.14	1.14	1.15	1.16	1.17
0.25	1.01	1.02	1.03	1.03	1.03	1.03	1.04	1.04
0	1.00	1.00	1.00	1.00	1.00	1.00	1.01	1.01

The parameter t is determined according to the formula (5.17) – see Chapter 5.

According to (5.17), one can state that for $\Delta r \le \Delta a$ we have $t \le 0$, and then $K_{\sigma\lambda} = 1$, that is, there is no reduction of the gearing load-bearing capacity, for $\Delta r > \Delta a^0$ we obtain $t > 1$, and the coefficient $K_{\sigma\lambda}$ is increased so much (Table 15), that the gearing becomes inoperative.

Considering the non-linear relation between the load and contact stresses (see Chapter 11), we will determine the coefficient $K_\lambda = (K_{\sigma\lambda})^{1/0.687}$ according to the diagram (Fig. 24).

Therefore, when calculating t (5.17), the coefficient K_λ is approximately determined according to the diagram (Fig. 24), by means of this coefficient the gearing adaptability is taken into account in contact strength analysis.

The refined definition of the coefficient K_λ can be made according to the following relation

$$K_\lambda = 1 + y(1 - t^{0.7}\varepsilon)^x, \qquad (10.23)$$

where $y = 0.025C_{\alpha\beta}^{0.75}$ within the range $0.015 \le C_{\alpha\beta} \le 0.05$; $y = 0.009C_{\alpha\beta}^{0.4}$ within the range $0.05 < C_{\alpha\beta} \le 0.2$; $y = 0.018C_{\alpha\beta}^{0.82}$ within the range $0.2 < C_{\alpha\beta} \le 0.3$; $\varepsilon = 0.9174C_{\alpha\beta}^{0.6008}$ within the range $0.015 \le C_{\alpha\beta} \le 0.1$; $\varepsilon = 0.939C_{\alpha\beta}^{0.6109}$ within the range $0.1 < C_{\alpha\beta} \le 0.3$; $x = 5\,C_{\alpha\beta}^3 (5 - C_{\alpha\beta}^{-3.72})$ within the range $0.015 \le C_{\alpha\beta} \le 0.3$.

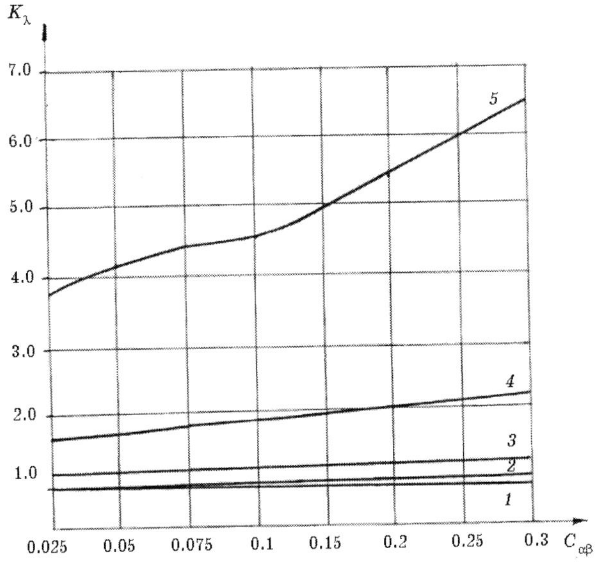

Figure 24. Diagram $K_\lambda = f(t, C_{\alpha\beta})$: lines $1 \ldots 5$ denote correspondingly $t = 0; 0.25; 0.5; 0.75; 1$.

10.6. Accounting the Dynamic Component of the Load, Appearing in Engagement

When estimating the dynamic load in strength analysis of NG, coefficients are often used, taken from the analysis of involute gearing [43], ignoring the differences between them. Investigation of complicated dynamic systems of NG [92, 209] have not yet led to engineering formulas acceptable for calculation.

The attempt is made below to consider the specific character of NG, on the basis of the described in [240] design scheme of the teeth edge impact, appearing due to errors of tooth engagement and deformations under load.

The impact force is determined according to the formula

$$U = v_y \sqrt{MC_n}, \tag{10.24}$$

where v_y is the impact velocity, *m/s*; *M* is the reduced mass of gears, *kg;* C_n is the rigidity of the tooth pair, *N/m*.

Analysis of kinematics of cylindrical NG TLA showed, that for the unfavorable combination of errors $F_{\beta r1}$, $F_{\beta r2}$ of pinion and gear tooth direction and deviations from axes parallelism f_{xr} and misalignment f_{yr}, the error of the angle of driven gear rotation for the uniform rotation of the driving pinion is increased practically linearly. Two characteristic phases are observed during the engagement of one pair of teeth: contacting at the dedendum and contacting at the addendum of the pinion tooth. The duration of these phases is proportional to the relation $\varepsilon_q/(1-\varepsilon_q)$, where ε_q is the phase axial overlap ratio (see Chapter 2). In the presence of errors, the kinematic (without load) contact of active flanks in every instant time takes place only at one contact point, and clearances are generated in the vicinity of other theoretical contact points. Assuming, on the basis of the described above, that the probable clearance Δ is equal to the product of the impact velocity and the duration of one phase of engagement, we will obtain [228]:

$$v_y = v\Delta \cos\beta / [\pi m(1-\varepsilon_q)], \tag{10.25}$$

where *v* is the tangential velocity, *m/s*.

Determining the clearance Δ according to probability methods [129], the rigidity C_n as the value, equal to the ratio of the normal force F_n (10.2) to the elastic displacement W_δ (10.11) and the reduced mass according to the formula [228]

$$M = 3.063 \cdot 10^{-6} b_w m^2 z_1^2 z_2^2 / [\cos^2\beta \cos^2\alpha_k (z_1^2 + z_2^2)], \tag{10.26}$$

we will determine the impact force *U* (10.24) and the coefficient, considering the preceding resonance dynamic load [189]

$$K_v = 1 + \varphi U / (K_A F_n). \tag{10.27}$$

Here φ is the coefficient, taking into account the influence of the dynamic load on endurance; basing on relations of [189], one can assume: for bending endurance tooth analysis $\varphi=1$; for analysis of flanks contact endurance $H_{HB}>HB\ 350$ and $H_{HB}\leq HB\ 350$ the values $\varphi=0.66$ and 0.33 correspondingly; K_A is the coefficient, taking into account the external dynamic load.

As an example (Table 16), letus give the calculation results of the dynamic load coefficient K_{Fv} for the bending endurance of NG according to the described technique and its involute analog according to [43]. Parameters of the gearing are: $z_1=16;\ z_2=43;\ m=4$ $mm;\ b_w=46\ mm;\ \beta=19°;$ the basic rack profile is taken according to the standard GOST 30224-96; flanks hardness is *HB 600*. It is accepted, that $T_2=1670\ N\cdot m$, $K_A=1$.

The comparison of coefficients K_v for Novikov and involute gearing, carried out in a wide range of geometrical parameters and loads, allowed to establish the following:

a) for general engineering gearing, operating at tangential velocities under *25 m/s* and having low degree of accuracy, the coefficient K_v for the NG is always less than for the involute one, though the difference is insignificant;
b) with the reduction of the gear manufacture accuracy, the difference in coefficients K_v for NG and involute one is increased for benefit of the former;
c) the relation of the fractional part (increment to unit) of coefficients K_v (10.27) in analysis of contact and bending endurance of soft and hard teeth correspondingly, which is equal to 1:2:3 for involute gearing [43], remains the same also for NG [228].

Table 16. The comparison of coefficients K_{Fv} for Novikov gearing (in numerator) and involute analogue (in denominator).

Degree of accuracy according to [40]	*Coefficient K_{Fv} for the speed v, m/s*			
	5	*10*	*15*	*20*
8	$\frac{1.030}{1.032}$	$\frac{1.060}{1.063}$	$\frac{1.090}{1.095}$	$\frac{1.120}{1.126}$
9	$\frac{1.036}{1.042}$	$\frac{1.072}{1.084}$	$\frac{1.108}{1.126}$	$\frac{1.144}{1.168}$
10	$\frac{1.046}{1.057}$	$\frac{1.092}{1.114}$	$\frac{1.138}{1.171}$	$\frac{1.184}{1.228}$

The described above allows for the calculation of the coefficient $(K_v)_{inv.}$ according to formulas [43] for involute gearing, not complicating the calculation and without great prejudice to the accuracy, by introducing a certain correction μ for NG, depending on the number *k* of the degree of accuracy according to [40]:

$$(K_v)_{Nov.}=[(K_v)_{inv.}-1]\mu+1, \tag{10.28}$$

$$\mu = 1 \text{ - for } k = 6, 7;$$
$$\mu = 4.23k^{-0.72} \text{ - for } k = 8 \ldots 12. \quad (10.29)$$

Chapter 11

SURFACE CONTACT STRENGTH ANALYSIS OF CYLINDRICAL NOVIKOV GEARING

11.1. CONTACT ENDURANCE ANALYSIS OF TOOTH ACTIVE FLANKS

Novikov gearing appeared first of all as a gearing with high contact endurance due to the elimination of the near-pitch-point zone from operation and possibility to control the curvatures of tooth active flanks [158].

Wide application of NG and practice of its operation showed that, in order to achieve the increased contact strength, it is necessary to have, on the one hand, sufficient height extension of active flanks and, on the other hand, the sensitivity to real errors, eliminating the edge contact and contact in the near-pitch-point zone.

Until 1 recently, two main approaches have been applied to surface contact endurance analysis of NG.

The first approach is based on the assumption that, during the tooth run-in process, the contacting of their convex-concave surfaces become linear in the direction, normal to the tooth line (that is, along its height) [158]. The well-known Hertz relation is used here for the calculation of normal stresses for the linear contact, the load is directly proportional here to the squared stress. This approach is developed in works of LMI [140] on the basis of the carried- out series of tests of NG with low hardness with the basic rack profile according to [45] and in works of ZAFEA [172].

The analysis of this approach [239] allows to make the following considerations. The calculation formula, proposed in [140], has the limited semi-empirical character, since it was obtained for gears with tooth hardness $H_{HB} \leq HB\ 320$ with the basic rack profile according to [45] and it includes a number of coefficients, "adjusting" the formula with respect to results of testing. That is why it can't be universal for NG calculation.

Design relations, pretending to be universal and proposed in [172], also have a number of evident drawbacks. Firstly, an important factor of gearing sensitivity to errors is not considered there at all. Secondly, a circumstance that, under considerable loads, the process of active flanks contact failure can surpass the process of run-in and the beginning of linear contact is not taken into account This probably explains the introduction into calculation of special correction functions, obtained by experimental data processing. However, in spite of it, the proposed in [172] relations give evidently overestimated (sometimes up to *2...3* times) design load-bearing capacity of the gearing compared with the real one, especially for high tooth hardness and the low degree of gears manufacturing accuracy. Moreover, these relations do not take into account non-zero gearing [129].

And finally, the described approach ignores the evident fact that, in the presence of even slight gear rims radial run-out, the linear contact in the gearing is impossible. It is known [171] that even for low tooth hardness, after their run-in, a certain difference of convex and concave contacting profiles radii is observed, demonstrating the preservation of theoretically point character of tooth contacting. This is especially true for gearing with hard teeth, where the run-in is less intensive, which has been clearly stated during tests of nitro carburized pairs with the basic rack profile RGU-5 at Izhevsk "OOO "Reduktor" - the contact pattern did not reach the apex edge of teeth [22].

The second approach, which appeared after the first one, is based on the assumption of theoretically point contact of teeth, determined by the initial geometry of their active flanks [196]. In this case, not normal, but effective stresses are applied as design ones, determined on the basis of solving the Hertz task for the point contact of two surfaces of the second order and according to the octahedral hypothesis of strength [96]; thanks to it, there appears opportunity to use in calculations the allowable normal stresses at linear contact [43], since there is a relation between effective σ_{He} and normal σ_H stresses on the surface

$$\sigma_{He} = 0.4\sigma_H \; . \tag{11.1}$$

For the second approach, the load is proportional to the cubic stress.

Accepting the relevance of calculation according to effective stresses, it should be noted nevertheless, that the application of the initial geometry of tooth flanks do not consider at all the fixed phenomenon of the stable process of the natural run-in (at any case for tangential velocities under *20 m/s*) of convex-concave contacting surfaces in NG, as a consequence of it, with the increase of the number of operating cycles the initial geometry of surfaces in the contact area is changed and σ_{He} is decreased. That is why the contact endurance analysis without account of the run-in always leads to abruptly reduced load-bearing capacity of the gearing compared with the real one, which is especially distinct for teeth with $H_{HB} \leq HB\ 350$. In methodical recommendations [176], the attempt to compensate the noted reduction by the increase of the allowable stress by *40%* was made. However, this contradicts with the physical nature of the material operation and leads to unreasonable results of calculation.

Though the quality aspect of the tooth run-in phenomenon of NG mentioned by many investigators is of no doubt, the quantitative aspect of this process is almost unexplored. However, by introducing some conditional coefficient Z_l of completeness of the tooth run-in along the height, one can considerably approximate contact calculations to the truth, by matching them with experimental results and simultaneously leaving the possibility to refine this coefficient in further investigations.

In order to calculate the surface contact endurance, a scheme is further proposed where the calculation is performed according to effective stresses of a theoretically point Hertz contact with an account of run-in. It is assumed that the profile reduced radius of curvature ρ_α is increased until the minor axis of the contact ellipse reaches the height, equal to $Z_l K_l l$; the longitudinal radius of curvature ρ_β, determining the kinematics of gearing, remains constant and equal to the initial one. (Strictly speaking, the contact pattern is not an ellipse [99, 169], the ellipse is conditionally taken for simplifying the calculations). The initial geometry in the proposed approach is used only in calculation of effective stresses (contact strength) under the action of maximum load (see below).

According to the accepted scheme, the minor semi-axis b_H of the contact ellipse is

$$b_H = 0.5 Z_l K_l l^* m . \quad (11.2)$$

On the other hand, the value b_H of the minor semi-axis can be expressed by the formula [196]

$$b_H = \overline{\beta} n_e (F_{Hn} \rho_\beta^* m / E)^{1/3} , \quad (11.3)$$

where $\overline{\beta} = b_H / a_H$ is the ratio of the minor semi-axis b_H of the ellipse to the major semi-axis a_H (ellipticity coefficient); F_{Hn} is the normal calculation force; E is the modulus of elasticity; n_e is the coefficient, depending on the ratio $C_{\alpha\beta} = \rho_\alpha / \rho_\beta$ and is calculated by means of elliptical integrals. Approximately, it can be taken according to tables [96] or determined by the relation

$$n_e = 1.15 C_{\alpha\beta}^{-0.094} . \quad (11.4)$$

The effective stress and normal force are related by the dependence [96]:

$$\sigma_{He} = K_e [F_{Hn} (E / \rho_\beta)^2]^{1/3} , \quad (11.5)$$

where K_e is the coefficient, depending on $C_{\alpha\beta}$ and also calculated by means of elliptical integrals. Approximately, it can be taken according to tables [96] or determined by the formula

$$K_e = 0.074 C_{\alpha\beta}^{-0.57} . \quad (11.6)$$

(Here and further, only the case $\overline{\beta} \le 0.45$ is considered, occurring in practice, when the maximum effective stress appears in the center of the contact ellipse).

Considering the approximate relation

$$\overline{\beta} = C_{\alpha\beta}^{0.63} , \quad (11.7)$$

established by numerical simulation, and taking $E = 2 \cdot 15 \cdot 10^5$ MPa (for steel gears) and reducing the effective stress σ_{He} to the normal one according to (11.1), we'll obtain the calculation formula after transformations:

$$\sigma_H = Z_H [T_H / (z cos\alpha_k)]^{0.687} m^{-2.063} (\rho_\beta^*)^{-0.312} (Z_l K_l l^*)^{-1.063}, \tag{11.8}$$

where T_H is the design torque at the gear, $N \cdot m$; Z_H is the coefficient, taken to be equal to 3960 for gearing with the basic rack profiles, having $x_a=0$, and equal to 3700 for gearing with the basic rack profile, having $x_a>0$.

The height l^* at the tooth of the basic rack is determined according to the formula (3.18).

The value of the coefficient Z_l of run-in completeness with account of experience of testing the gearing with teeth of various hardness can be determined according to the relation:

$$Z_l = 0.95 - 0.00037(H_{HB} - 200), \tag{11.9}$$

where H_{HB} is the least hardness of tooth active flanks for the pinion and gear. For $H_{HB} \leq HB\ 200$ the value $Z_l = 0.95$ should be taken, and for $H_{HB} > HB\ 550$ the value $Z_l = 0.82$ should be taken.

As it was noted above, the coefficient K_l considers the reduction of the tooth active height due to tooth cutting by generation.

K_l is determined according to the relation [134, 135]:

$$K_l = min\{K_l^{(1)}, K_l^{(2)}\} \cdot 10^{-3}, \tag{11.10}$$

if $K_l > 1$ here, it is taken $K_l = 1$.

In its turn with account of a certain correction, depending on the module:

$$K_l^{(1)} = min\{K_{KA} K_{A1}, K_{KI} K_{I2}\} + min\{K_{KP} K_{P1}, K_{KF} K_{F2}\}, \tag{11.11}$$

$$K_l^{(2)} = min\{K_{KA} K_{A2}, K_{KI} K_{I1}\} + min\{K_{KP} K_{P2}, K_{KF} K_{F1}\}. \tag{11.12}$$

Coefficients K_A, K_P, K_F and K_I depend on the reduced number $z_v = z / cos^3 \beta$ of gear teeth and on the shift coefficient x^* of the gear basic rack profile, they are taken according to Tables 1-3 of the Appendix 2 for gears, manufactured with various basic rack profiles.

The correction coefficients K_{KA}, K_{KP}, K_{KF} and K_{KI}, depending on the module, are taken according to Table 17.

The condition of the surface contact strength is written in the form

$$\sigma_H \leq \sigma_{HP}. \tag{11.13}$$

Table 17. Correction coefficients K_{KA}, K_{KP}, K_{KF}, K_{KI}.

Coefficient	Basic rack profile						
	RGU-5						
	m, mm						
	≤ 2.5	Over 2.5 to 3.15	Over 3.15 to 4.0	Over 4.0 to 5.0	Over 5.0 to 6.3	Over 6.3 to 8.0	>8
K_{KA}	1.022	1	1	1.022	1	1	1.022
K_{KP}	0.98	1	1	0.98	1	1	0.98
K_{KF}	1.01	1	1.005	1.025	0.99	0.98	1.015
K_{KI}	1.014	1	1	1.023	1.011	1.015	1.02
Coefficient	According to the standard GOST 15023-76				According to the standard GOST 30224-96		
	m, mm				m, mm		
	≤ 3.15	Over 3.15 to 6.3	Over 6.3 to 10	>10	≤ 4.5	Over 4.5 to 9	>9
K_{KA}	1.015	1	1.015	1.018	1	1	1
K_{KP}	0.98	1	0.98	0.97	1	1	1
K_{KF}	0.97	1	0.95	0.93	1	0.96	0.92
K_{KI}	1	1	1.052	1.085	1	1.01	1.02

The formula (11.8) allows for the use of the value of the allowable stress σ_{HP} (separately for the pinion and the gear) from [43]:

$$\sigma_{HP} = \sigma_{Hlim} Z_N Z_R Z_v Z_L Z_X / S_H . \qquad (11.14)$$

Here the designations are:

σ_{Hlim} is the contact endurance limit of tooth active flanks, corresponding to the basic number of stress cycles alternation, it depends on the method of heat and chemically heat treatment of teeth and hardness of their flanks, it is taken from [43]. The rest component coefficients of (11.14) are also taken from [43] and they denote:

Z_N is the durability coefficient, equal to

$$Z_N = \sqrt[q_H]{N_{H\,lim} / N_H} , \qquad (11.15)$$

for $N_H \le N_{Hlim}$ $q_H = 6$, and for $N_H > N_{Hlim}$ $q_H = 20$; the value Z_N must not exceed *2.6* for a homogenous structure of material and *1.8* for case-hardening, and also must not be under *0.75*. (When applying the method of equivalent cycles, N_{HE} is substituted instead of N_H); N_{Hlim} is the basic number of stress cycles alternation, equal to

$$N_{Hlim} = 30H_{HB}^{2.4} \le 120 \cdot 10^6 ; \tag{11.16}$$

Z_R is the coefficient, accounting for the roughness of tooth conjugate flanks, for well run-in NG it is taken $Z_R=1$; Z_v is the coefficient, accounting the tangential velocity, equal to

$$Z_v = 0.85v^{0.1} \text{ for } H_{HV} \le HV\,350 ; Z_v = 0.925v^{0.05} \text{ for } H_{HV} > HV\,350 ; \tag{11.17}$$

Z_L is the coefficient, accounting the influence of lubrication, it is taken $Z_L=1$; Z_X is the coefficient, accounting the dimension of the gear, equal to

$$Z_X = \sqrt{1.07 - 10^{-4} d} , \tag{11.18}$$

for $d < 700mm$ it is taken $Z_X=1$; S_H is the strength safety factor, see Chapter 14 for more details about S_H.

11.2. Contact Strength Analysis of Tooth Active Flanks under the Action of Maximum Load

As it was noted above for this analysis, effective stresses and initial geometry of teeth are applied, that is, the process of run-in is not taken into account.

The strength condition has the form

$$\sigma_{Hmax} \le \sigma_{HPmax}, \tag{11.19}$$

where the allowable contact stress σ_{HPmax} under the maximum load T_{Hmax} is taken according to [43] depending on the heat and chemically heat treatment of teeth, and also on their surface hardness.

Letus give the initial relations and a brief derivation of the calculation formula for determination of σ_{Hmax}.

The effective stress in the center of the contact ellipse is expressed by a well-known relation [196]

$$\sigma_{Hemax} = (1 - 2\nu) Z_{HM} \sigma_{kmax} \tag{11.20}$$

where ν is the Poisson's ratio for gear materials; Z_{HM} is the coefficient, accounting the shape of the contact area and equal to

$$Z_{HM} = (1 - \bar{\beta} + \bar{\beta}^2)^{0.5} / (1 + \bar{\beta}), \tag{11.21}$$

σ_{kmax} is the maximum normal stress in the center of the contact ellipse

$$\sigma_{k\,max} = 1.5F_{Hn}/(\pi a_H b_H). \tag{11.22}$$

Reducing the effective stress to the normal one according to (11.1) and taking $\nu = 0.3$ for steel gears, we'll obtain

$$\sigma_{H\,max} = Z_{HM}\sigma_{k\,max} \tag{11.23}$$

Let's express the major a_H and minor b_H semi-axes of the contact ellipse by well-known formulas:

$$a_H = n_a[1.5\eta F_{Hn}\rho_\alpha^* m/(1+C_{\alpha\beta})]^{1/3}, \tag{11.24}$$

$$b_H = n_b[1.5\eta F_{Hn}\rho_\alpha^* m/(1+C_{\alpha\beta})]^{1/3}. \tag{11.25}$$

In equations (11.24) and (11.25) n_a, n_b are coefficients, determined by the geometry of contacting surfaces and taken approximately according to the relation [176]:

for $0.01 \le C_{\alpha\beta} \le 0.1$: $\quad n_a = 0.94943C_{\alpha\beta}^{-0.39947}, \quad n_b = 0.86244C_{\alpha\beta}^{0.21033};$ (11.26)

for $0.1 \le C_{\alpha\beta} \le 0.4$: $\quad n_a = 0.97016C_{\alpha\beta}^{-0.39010}, \quad n_b = 0.9509C_{\alpha\beta}^{0.25280};$ (11.27)

η is the coefficient, accounting the elastic constants of gear materials, $\eta = (1-\nu_1^2)/E_1 + (1-\nu_2^2)/E_2$.

Accepting $\nu_1 = \nu_2 = 0.3$ and $E_1 = E_2 = 2.15\cdot 10^5$ MPa, we will obtain

$$\eta = 8.47\cdot 10^{-6}, \ MPa^{-1}. \tag{11.28}$$

Substituting the values from (11.22), (11.24), (11.25), (11.28) into (11.23) and passing from the calculation force to the calculation torque, we will finally obtain:

$$\sigma_{H\,max} = 11050Z_{HM}\{T_{H\,max}(1+C_{\alpha\beta})^2/[z\cos\alpha_k(\rho_\alpha^*)^2]\}^{1/3}/(n_a n_b m). \tag{11.29}$$

In accepted designations, the coefficient $\overline{\beta}$ of the ratio of the minor semi-axis of the contact ellipse to the major one, which is the component of (11.21), is equal to n_b/n_a.

Chapter 12

Depth Contact Strength Estimation of Novikov Gearing

Depth contact failures (DCF) are the type of failure, specific for high-stressed, mainly heat or chemically heat strengthened parts (see Fig. 18). With respect to surface failures, DCF is an independent type of failure with a principally different mechanism of contact-fatigue durability. If in the case of surface incipient cracks, which are considered to be present initially, their source is the inevitable surface defects (including traces of tool machining) and durability is determined by intensity of cracks growth [78, 217, 269], then in the case of DCF, the main part of the lifetime falls to the stage, preceding the appearance of a crack.

Both depth and surface processes run simultaneously, and the leading type of failure is determined by a complex of factors, the main of them is the stress level and properties of material near-surface layers. The scheme of contact-fatigue durability when applied to gearing is shown in Fig. 25, where N_H is the number of loading cycles, σ_{kmax} is the maximum contact stress on the surface. Fig. 25 shows that DCF are the limiting factors mainly in the area of the restricted durability.

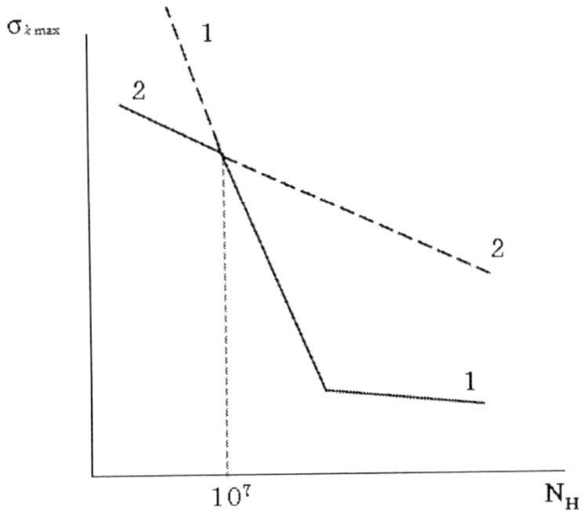

Figure 25. Typical diagram of the contact-fatigue durability of surface hardened gearing: 1–1 – for surface failures; 2–2 – for depth failures.

The main manufacturing technique of increasing the level of the depth contact strength (DCS) is the increase of the surface hardened layer thickness, which cannot be unlimited for a gearing since, despite economic considerations, the heat and chemically heat interaction causes the reduction of material plastic properties, which has a negative effect on tooth fracture strength. Compared with the DCS, as a rule, the surface hardened layer of smaller value is optimal according to the fracture strength.

In involute gearing, the risky zone according to the fracture is within the fillet, that is, outside the active segment, which principally allows to vary the thickness of the hardened layer along the tooth height. In NG, the cases are possible when the dangerous zone according to the fracture covers active profile segments of the dedendum.

It is known that, for conventional involute gearing the reasonable parameters of the hardened layer are determined (mainly experimentally), the main characteristics of the layer – its thickness – being rated according to the gearing module. However, application of these recommendations for NG without the thorough analysis is wrong since, in this type of gearing, the radii of contacting flanks curvature, determining the stressed state in the zone of contact, are the parameters of the basic rack profile, proportional to the module. The most widespread basic rack profiles (according to the standard GOST 15023-76, DON-63, DLZ-0.7-0.15, DLZ-1.0-0.15, RGU-5, according to the standard GOST 30224-96 and oth.) are considerably different according to these parameters, that is why the stressed state in the contact zone will differ even for NG with different basic rack profiles.

Experimental determination of hardened layer reasonable parameters without corresponding theoretical ground is a long-term and expensive process. Hence, the urgency appears of predicting the level of DCS on the basis of, at least, phenomenological approaches.

12.1. Existing Methods of Depth Contact Strength Estimation for Surface Hardened Gearing

The conditions of DCF appearance have been investigated by a number of authors. Design models, applied here (their summary is given below in Table 18), can be divided into two groups. For the first one (models *NN1-6*), the goal is stated to give a quantitative estimation of the strength safety factor. For the second one (model *N7*), they confine themselves to determine qualitatively the potentially risky zones of the hardened layer – the so-called "risky areas". The common for all models is the postulation of the linear dependence between the strength of material and its hardness for the whole layer. However, application of different hardness scales here demonstrates the absence of the common viewpoint on the hardness as the sufficient measure of the material strength, especially for $H_{HV} > HV\ (450...500)$.

It is designated in Table 18 and further: σ_i is the intensity of octahedral stresses; τ_{max} is the maximum (direct) tangential stress; τ_{yz} is the orthogonal (alternating-sign) tangential stress; σ_1, σ_2, σ_2 are the main stresses; H_{100} is the micro hardness; k is a certain coefficient, taking into account the degree of normal stresses influence on the effective ones; x, y are coordinates in the plane, perpendicular to the axis z, coinciding with the common normal to contacting flanks at the theoretical point of contact.

Table 18. Applied methods of depth contact strength evaluation.

№	Source	Strength criterion according to	Design effective stress σ_{HKe}	Limiting effective stress σ_{HKlime}	Notes
1	[32]	Mises	$0.5\sigma_i$	0.84 H_{HB} 0.84 H_{HV}	$H_{HB} \le HB350$ $H_{HB} > HB350$
2	[107, 196]				For Novikov gearing
3	[31]	Gest-Moor	$\tau_{yz} - 0.2\sigma_y$	1.07 H_{HB}	
4	[61]		$\tau_{yz} - k\lvert\sigma_y\rvert$	0.8 H_{100}	$k = 0.25$ for $H \le 600H_{100}$, $k = 0.40$ for $H > 600H_{100}$
5	[218]	Tresk	$0.5(\sigma_1 - \sigma_3)$	1.00 H_{100}	
6	[258, 268]			1.00 H_{HB}	
7	[226, 269]		$max(\tau_{max}/H_{HV})$ $max(\sigma_i/H_{HV})$ $max(\tau_{yz}/H_{HV})$		

When applying the models *NN1, 2, 3*, the hardness of the core and adjacent to it, part of the layer (the so-called sublayer) are mainly determined, with application of relations of mechanical characteristics of medium viscosity steels with the hardness under *HB 350*. The field of their reasonable application is the chemical heat treatment types, under comparatively thin diffusion layers (nitriding) are generated. Processes, running in the hardened layer itself, are shown insufficiently correctly.

In the model *N4*, elements of which are used in [43], not only medium-hardness layers are applied, but the difference of strength factors is also considered by the coefficient k. However, for $k = 0.4$ (for layers with $H > 600\ H_{100}$) for the plane deformation at depths under $0.4\,b_H$ (where b_H is the semi-width of the band under linear contact or the minor semi-axis of the ellipse for the theoretical point contact) the effective stress $\sigma_e < 0$, and the model is inapplicable. Moreover, due to discreteness of the coefficient k in the area $H = 600\ H_{100}$, the "burst" of strength factors takes place, contradictory to logics. The technique, recommended in [43], is directed towards the evaluation of DCS within the zone of maximum effective stresses action at the depth $z = (0.6...0.8)\ b_H$ and it "does not catch" experimentally visible sites of DCF at other depth levels.

On the whole, most of the models are based on yield criteria and they only note the possibility of plastic deformations appearance. Actually, micro plastic phenomena in the contact area are concentrated mainly at the depth of maximum direct shearing stresses action. However, the question if these changes are the measures of the contact fatigue exactly, has not been solved yet [225].

The role of shearing stresses in the failure process is reduced with the growth of material brittleness, which is observed for bearing and tool steels after quenching. And if there is a close to linear relation $\sigma_{vr+} \approx 0.34H_{HB}$ for steels with hardness $H_{HB} < HB\ (450...500)$, and the failure has the viscous nature, then further increase of hardness leads to violation of this relation. Failure will be brittle or mixed, depending on the loading rigidity. The differences in strength limits are distinctively shown up for tension and compression: $\sigma_{0,2+} \approx \sigma_{vr+}$, $\sigma_{0,2-} << \sigma_{vr-}$, where $\sigma_{0,2+}$, σ_{vr+} are the conventional yield point and ultimate strength for the tension correspondingly, $\sigma_{0,2-}$, σ_{vr-} are the same for compression. Hardness is not the sufficient factor of the material strength any more – the additional account of its properties is necessary of plastic ones [12, 33, 178]. The noted factors are of peculiar importance for the estimation of the load-bearing capacity of the diffusion layer with the pronounced non-homogeneity of structure: from hypoeutectoid (in the zone of transfer to the core) to a hypereutectoid one (in near-surface areas).

All of the enumerated above defines the high degree of the calculation conditional character according to the known models and, as a consequence, the necessity to introduce considerable (up to *1.5* for stresses) calculation strength safety factors. Moreover, the described models do not allow the evaluation of a well-known in practice difference of gearing (and other parts) operation characteristics, subjected to carburizing and nitro carburizing. That is why it is necessary to create a calculation model, which would allow the prediction of the appearance of "risky areas" along the whole thickness of the diffusion layer, and moreover, consider the difference between strength properties of the material depending on the type of chemically heat treatment.

12.2. Criterion of Pisarenko-Lebedev and Terms of Its Application when Calculating DCS

Failure of material in general cases is determined by its ability to resist both tangential and normal stresses. The most promising, with respect to considerable structural non-homogeneity of the material,is the application of generalized criteria of the limiting state, in particular, the criterion of Pisarenko-Lebedev for structurally non-homogeneous material [162, 178]:

$$\sigma_e = \chi\sigma_i + (1-\chi)\sigma_1 A^{1-(\sigma_1+\sigma_2+\sigma_3)/\sigma_i} \le \sigma_{e+}, \tag{12.1}$$

where σ_{e+} is the limiting effective stress for $\sigma_1 > 0$; $\chi = \sigma_+/\sigma_-$ is the parameter of the material plasticity, considering the degree of shearing deformations participation in micro failures; σ_+, σ_- are breakage (failure) stresses for the uniaxial tension and compression correspondingly; A is the static parameter of the defectiveness, its value for hardened steels is $A = 0.7...0.8$.

Note, that the criterion (12.1) is universal, that is, it can be applied both for linear and theoretical point contacting, that is why its grounding will be based on experiments both for cylindrical rollers and involute gearing, and for NG (see below).

For plastic materials $\sigma_+ \approx \sigma_-$, $\chi = 1$, and the formula (12.1), represents the Mises criterion. For absolutely brittle materials $\chi \to 0$, and the transition to the criterion of maximum normal stresses takes place.

Considering that the criterion (12.1) is solved with respect to tensile stresses, the power exponent for the defectiveness parameter must be taken with respect to the module (that is, according to the absolute value), and limiting values of effective stresses can be presented as:

$$\sigma_{e+} = \chi \sigma_{HKP} k_e, \qquad (12.2)$$

where $k_e = \sigma_{emax}/\sigma_{kmax}$ is the equivalence coefficient.

Obviously, both parts of (12.2) are in the functional dependence on the parameter χ.

In order to develop the design model, it is necessary to investigate the influence of the parameter χ on the stressed state, and also to have information on specific values of this parameter. For this purpose, letus formulate the main backgrounds and assumptions:

1) the criterion (12.1) is a criterion of the static strength; the task is considered in the quasistatic statement on the basis of the well-known correlation of strength and fatigue characteristics [178];
2) properties of the diffusion layer areas with hypereutectoid and eutectoid structures are taken similar to properties of hardened bearing and tool steels, and of areas with hypoeutectoid structure – to properties of thermally improved or hardened medium-carbon structural steels;
3) the influence of residual stresses is assumed to be shown up in mechanical characteristics of steels after the corresponding heat treatment [66, 211];
4) mechanical properties of the material, including plasticity characteristics, are changed along the thickness of the diffusion layer and considered as the function of hardness;
5) the contact is assumed to be elastic, with constant elastic characteristics of the material (independently on the chemical composition, structure and hardness).

The criterion (12.1) takes into account the character of the stressed state and the difference between material resistance to tension and compression, characterized by the plasticity parameter. It is required here to estimate the influence of this parameter on the strength level depending on its value and contacting conditions. For this purpose, the analysis of effective stresses distribution along the depth within the semi-width of the contact area was carried out with variation of the parameter χ from *1.0* to *0.4* and the ellipticity coefficient $\bar{\beta} = b_H / a_H$ from *0* to *0.9*.

The influence of the parameter χ on the strength level was investigated by the ratio of allowable stresses to the acting effective ones, that is, by the design safety factor S_{HK}:

$$S_{HK} = \frac{\chi \sigma_{HKP} k_e}{\chi \sigma_i + (1-\chi)\sigma_1 A^{\left|1-(\sigma_1+\sigma_2+\sigma_3)/\sigma_i\right|}}, \qquad (12.3)$$

where σ_{HKP} is the allowable depth normal contact stress.

Investigation of the plasticity parameter influence on the stressed state for cases of a linear and a theoretically point contact showed:

a) effective stresses achieve their maximum values in the center of the contact area, and the less the value of the parameter χ (Table 19) is, the deeper the absolute maximums are arranged with respect to corresponding maximums of intensity of octahedral stresses;
b) the reduction of χ leads to extension of the area of the increased relative stress level, expressed by the ratio of the current depth contact stress σ^z_{HKe} to the maximum one $(\sigma_{HKe})_{max}$ (Fig. 26);
c) for the linear and point (up to $\overline{\beta} = 0.45$) character of contacting the equivalence coefficient k_e is satisfactorily described by an approximating relation:

$$k_e = 0.62574(\chi - 0.11128); \qquad (12.4)$$

d) the function (12.3) decreases in the area of small values of χ, and, in all cases, it is not less than the value $\chi = 0.35$, peculiar for cast irons; for greater values of χ, typical of steels, the increase of the plasticity parameter (for invariable other conditions) leads to the increase of strength.

Table 19. Maximum rated effective stresses $(\sigma^0_{HKe})_{max} = (\sigma_{HKe})_{max} / \sigma_{k\,max}$ (in numerator) and relative depth $z^0 = z / b_H$ of their occurrence (in denominator).

$\chi \rightarrow$ / $\overline{\beta} \downarrow$	0.4	0.5	0.6	0.7	0.8	0.9	1
0	*0.1829* / *1.000*	*0.2429* / *0.900*	*0.3035* / *0.800*	*0.3669* / *0.800*	*0.4296* / *0.750*	*0.4935* / *0.700*	*0.5575* / *0.700*
0.1	*0.1957* / *1.000*	*0.2560* / *0.900*	*0.3178* / *0.850*	*0.3806* / *0.800*	*0.4440* / *0.750*	*0.5079* / *0.750*	*0.5720* / *0.700*
0.3	*0.2084* / *0.900*	*0.2708* / *0.850*	*0.3347* / *0.800*	*0.3995* / *0.750*	*0.4648* / *0.700*	*0.5308* / *0.700*	*0.5969* / *0.700*
0.5	*0.2119* / *0.800*	*0.2761* / *0.750*	*0.3417* / *0.750*	*0.4081* / *0.700*	*0.4754* / *0.700*	*0.5428* / *0.700*	*0.6105* / *0.650*
0.7	*0.2128* / *0.750*	*0.2765* / *0.700*	*0.3434* / *0.650*	*0.4112* / *0.650*	*0.4795* / *0.600*	*0.5480* / *0.550*	*0.6173* / *0.550*
0.9	*0.2085* / *0.650*	*0.2747* / *0.600*	*0.3422* / *0.575*	*0.4111* / *0.550*	*0.4801* / *0.550*	*0.5497* / *0.500*	*0.6198* / *0.500*

Letus consider in more detail, the plasticity parameter.

The application of the criterion (12.1) is possible if the information on the value and character of variation of the plasticity parameter χ within the diffusion layer is available. However, the compression ultimate strength, necessary to determine χ, is not a regulated value. Informationavailable to the authors about it (Table 20), is fragmentary and often contradictory, which can be explained by the absence of the unified technique for its determination.

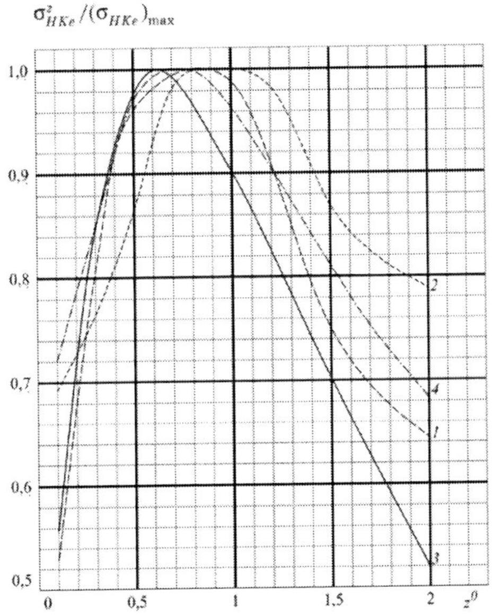

Figure 26. Distribution of effective stresses through the depth depending on the plasticity parameter: 1 – $\chi = 1.0,\ \bar{\beta} = 0$; 2 – $\chi = 0.6, \bar{\beta} = 0$; 3 – $\chi = 1.0, \bar{\beta} = 0.5$; 4 – $\chi = 0.6, \bar{\beta} = 0.5$.

Table 20. Mechanical characteristics of hardened steels.

Steel grade	σ_{vr+}, MPa	σ_{vr-}, MPa	$\tau_k^{*)}$, MPa	H_{HRC}	χ	Source
R–18	1900–2000	3400–3500	2250	62–64	0.58–0.62	[33]
H	1800	3450	2350	62–64	0.51	
R–18	1980	4100	1720	62–64	0.48	[186]
R–9	2160	4540	1850	62–64	0.42	
9HS	2130	5100	1830	62–64	0.42	
U12	2160	5150	1790	62–64	0.41	
40H	1580	3150	1375	45	0.50	
30HGS	1720	2210	–	–	0.78	
HVG	1600–1800	–	–	–	0.42–0.46	
ShH–15	1903	2943	2256	61–62	0.65	[166]
75HGST	–	–	–	48	1.00	
9HS	–	–	–	62	0.63	[205]
9HS	–	–	–	50	0.71	
ShH –15	2100	3000	–	58	0.70	[186]
ShH –15	1900–1940	2400	–	52	0.80	

*) τ_k is the shearing ultimate strength.

Letus establish in the first approximation the possible dependence of χ on the most easily controlled steel characteristics – hardness, with account (qualitative) of the influence of main elements of the hardened layer microstructure – martensite and retained austenite. For this purpose, letus consider the safety factor S_{HK} in the assumption of $\sigma_I = 0$ (the most unfavorable case, when the "relief effect" is absent): $S_{HK} = \sigma_{HKP} k_e / \sigma_i$. The increase of hardness finally leads to the strength increase (in any case, under $H_{HRC} = HRC\ 58$ [216, 217]), however, the plasticity parameter is decreased simultaneously, which, as it was demonstrated above, reduces the relative strength. Opposite influence of these factors allows to establish the level of the strength balance, for which $\frac{dS_{HK}}{dH_{HV}} = 0$ and $S_{HK} = const$.

On the basis of the experimental data processing and analysis, letus write the limiting normal stresses as [66, 218, 257]:

$$\sigma_{HKP} = c H_{HV}, \tag{12.5}$$

where c is a certain proportionality coefficient, and assign the boundary values of χ for steels, which do not contain nitrogen and nickel: $\chi_{H_{HV} < HV\,400} = 1.00$ and $\chi_{H_{HV} = HV\,850} = 0.6$. Let us determine the intermediate values of $(H_{HRC} \approx HRC\ 50)$ by comparing the conditional yield points, since the range $HRC45 < H_{HRC} < HRC52$ is characterized by more or less mixed properties. In this case , $\chi \approx 0.82...0.88$

The most important quality feature of austenite-martensite structures is the grain dimension. Its growth, in other equal conditions, increases the tendency of the material to brittle failure, that is, to reduction of the plasticity parameter value.

Contact durability in many respects is determined by the contents of the retained austenite A_u in the structure, expressed in %. In spite of the fact that A_u usually acts as the structure component, reducing the ultimate strength and yield point of steel [33], and most of standards restrict its contents in the layer by *(25...30)%* for carburizing and by *(40...50)%* for nitro carburizing, nowadays it became reasonable to increase its contents in the hardened layer up to *50%* for carburizing and *60%* for nitro carburizing. This contradiction can be explained as follows: the influence of A_u is revealed, in particular, by varying the power exponent of the limited durability [210]. The growth of the A_u content leads to the durability reduction under loads, close to limiting, but below the definite level the situation is quite opposite. For the tested steels, this level is within the range $\sigma_{k\,max} = 1700...2200$ *MPa.*

Comparison of carburizing and nitro carburizing shows that nitro carburizing provides, as a rule, a higher level of the load-bearing capacity. In many respects it is explained by the character of nitrogen in nitro carburized layers, where it is present in the solid solution, nitrides (carbo nitrides) and in the molecular form. But if the presence of the molecular nitrogen is a negative factor, showing the microstructure defectiveness, and its contents in the layer must be minimum, then in the solid solution the role of nitrogen is similar to the role of carbon, and increasing the nitride phase leads to the strength increase [66]. When the hardness is the same and constant along the thickness of the specimen (steel 12H2N4A), the value σ_{vr+} of specimens, subjected to nitro carburizing, is higher than that of carburized ones by

(10...12)% [136]. On the basis of these considerations, for small-grained and nitrogen-containing steels the value of the plasticity parameter is taken as *0.68...0.70* for the hardness *HV 850* (since the contact endurance limit is not increased, and σ_{vr+} is increased).

The simplest dependence within the range $HV400 \le H_{HV} \le HV850$, satisfying the stated requirements, is the linear one:

$$\chi = \begin{cases} -0{,}89\, H_{HV} \cdot 10^{-3} + 1.356, & (\chi_{min} = 0.6); \\ -0{,}71\, H_{HV} \cdot 10^{-3} + 1.284, & (\chi_{min} = 0.68). \end{cases} \quad (12.6)$$

12.3. Allowable Stresses

Fig. 27 shows well-known recommendations on the choice of ultimate and allowable compression stresses, applied in contact strength analysis of steel parts in case of linear contact. For theoretical point contact, the corresponding values are increased by *(30...50)%* [166], which is proved by practice. Consistent with the difference between shearing stresses at the contact surface, such recommendations contradict the conditions of the depth strength, since maximum shearing stresses are higher for the point contact. The reason of the contradiction is the discrepancy in stresses – design (under plane deformation) and actual, specified by the finiteness of real solids dimensions [212], and also the positive influence (for locally applied load compared with the distributed one) of neighboring less loaded areas of the material [178].

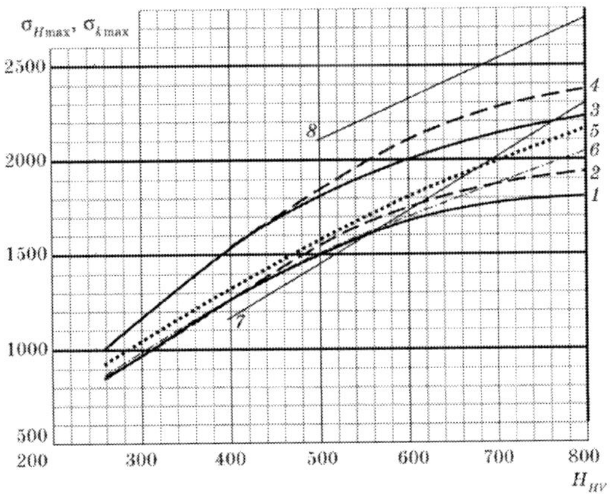

Figure 27. Ultimate compression stresses: 1, 2 – according to the proposed model for the linear contact, correspondingly for $\chi_{min} = 0.60$ and for $\chi_{min} = 0.68$ and the basic number of loading cycles $N_{HK\,lim} = 10^7$; 3, 4 – the same for a theoretical point contact; 5 – the limit of surface contact endurance according to [43] for $N_H = 10^7$; 6 – the limit of depth contact endurance according to [258, 268] ($\sigma_{HK\,lim} = 3.3 H_{HB}$); 7 – the limit of surface contact endurance according to [200] ($\sigma_{HK\,lim} = 2.887 H_{HV}$); 8 – $\sigma_{HP\,max} = HRC\ 44$ according to [43].

The level of allowable stresses according to DCS, has been determined on the basis of processing the experimental data [67, 210, 226, 257], reduced to the unified basic number of cycles - 10^7. Since DCF are mainly revealed in the area of the small-cycle and limited durability, the maximum normal compression surface stresses in this area must not exceed the corresponding stresses, allowable according to the surface contact strength.

Since the main bulk of experiments relates to cases of the linear contact, and the design model is invariant with respect to contacting conditions, the obtained results cover the cases of the theoretical point contact according to the given above recommendation.

The value of the coefficient c (12.5) is determined in terms of the equality $\sigma_{HKe} = \sigma_{HKPe}$ at the depth of incipient cracks with reduction to the unified base of 10^7 cycles. The index of the inclination angle of the fatigue curve is known only from experiments and [198, 226] and is characterized by a considerable scatter – from *8* to *28*. In [31, 32], it is recommended to assign it within the range *16...18*. We accepted in calculations the value, equal to *20*.

Design values of the coefficient c are within the range *3.23...3.40* for cases of DCF, both in near-surface and in conversion zones, for the probability of non-failure *0.95* and for the presence of only one potentially risky zone for the linear contact. It should be noted that minimum values of c accurately correspond to cases of cracks development at several levels of the depth.

Hence, one can write for the presence of one potentially risky zone within the hardened layer, introducing a number of coefficients, considering the factors, which influence the contact load-bearing capacity greatly, but are not related to the design model, on the basis of (12.1) - (12.3):

$$\sigma_{HKPe} = \chi(\chi - 0.11128) H_{HV} \cdot Z_{LK} \cdot \prod_{i=1}^{6} K_i \cdot \qquad (12.7)$$

The durability factor

$$Z_{LK} = (10^7 / N_{EK})^{1/20}, \qquad (12.8)$$

where N_{EK} is the equivalent number of stress cycles alternation.

The durability factor is within the range $(Z_{LK})_{max} = 1.125$ for not run-in NG (with original contact geometry), calculated at small number of cycles, to the value $(Z_{LK})_{min} = 1$ for run-in NG, calculated for the basic number of cycles. (The upper limit is according to the condition of non-exceeding the allowable depth stresses with respect to allowable surface ones).

The coefficient K_1 depends on the character of contacting. With the account of recommended (see above) relations of allowable stresses for the linear and point contact: $K_1 = 2.05...2.15$ - linear contact; $K_1 = 2.55...2.70$ - point contact.

The coefficient K_2 depends on the number of "risky areas" within the hardened layer, if there is one area, then $K_2 = 1$, if there are two areas - $K_2 = 0.90...0.95$; it is desirable to provide in gearing design that the parameters of the hardened layer would eliminate the possibility of the second "risky area" appearance.

The influence of the external tangential load is taken into account by the coefficient K_3, defined according to the relation [116]:

$$K_3 = \left\{1 + [(1 + 60C_{\alpha\beta}^{0.25} f^2)^{0.5} - 1](1 + z^0)^{-8.5}\right\}^{-1}, \tag{12.9}$$

where f is the factor of friction on the surface.

K_4 is the coefficient, taking into account the error of approximation of NG real flanks by surfaces of the second order and it is taken to be equal to 0.98 according to [99].

K_5 is the coefficient, taking into account the quality of the material and CHT, it is taken as: $K_5 = 0.90...0.95$ for carbon and low-alloy steels; $K_5 = 0.95...1.05$ for alloy steels with nickel contents under 1%; $K_5 = 1.05...1.10$ for chromium-nickel steels.

K_6 is the coefficient, taking into account the scatter of material mechanical characteristics in the hardened layer, equal to $0.90...0.95$ (smaller values – when the automated regulation of the process of CHT is absent).

12.4. CHECK-UP OF THE PROPOSED MODEL

In order to check the correctness of the proposed model, the estimation has been performed with respect to its conformance to the well-known experimental data and alternative design methods. The following methods have been analyzed as alternative:

1) qualitative – according to criteria of local maximums (Table 18, model *N6*);
2) quantitative – according to the criterion of maximum tangential stresses (Table 18, model *N5*) and recommendations [43].

Altogether, about 60 experiments were processed. As it was noted above, the proposed model was invariant with respect to gearing systems, in order to check its correctness, the experimental results were used for both linear contact (linear rollers, involute gears) and theoretical point contact (NG). The most characteristic results are shown in Table 21 and Fig. 28-31, where the designations are:

1) distribution of the hardness along the thickness of the diffusion layer;
2) S_{HK} according to the formula (12.3) for χ according to the formula (12.6) for $\chi_{min} = 0.6$ for boundary values of the coefficient K_1 (if only one line is present on the diagram – for the lower boundary);
3) the same for $\chi_{min} = 0.68$;
4) S_{HK} according to the Tresk criterion (Table 18) for $\sigma_{HKlime} = 1.00H_{HB}$ [258, 268];
5) σ_i / H_{HV}.

Table 21. Results of calculations and experimental investigations.

№	Specimen	Type of CHT	$\sigma_{k\max}$, *MPa*	N_{EK}	b_H, *mm*	Type of failure
1	Roller CC-60	C	1717	8.2 10^6	0.454	DCF in the effective zone [226]
2	Roller CD-30		1766	1.50 10^7	0.233	
3	Roller CA-120		1766	1.70 10^5	0.933	DCF in the sublayer [226]
4	Roller CB-120		1864	3.00 10^6	0.981	DCF at two levels [226]
5	Roller CB-30		1962	2.90 10^5	0.259	
6	Involute pinion, m = 4.0 *mm*	C	1900	1.60 10^7	0.280	DCF in the effective zone [257]
7	Involute pinion, m = 6.5 *mm*	C	1750	2.64 10^6	0.404	DCF in the effective zone and in the sublayer [210]
8		NC		9.24 10^6		Pitting [210]
9	Involute pinion, m = 6.0 *mm*	NC	1723	9.82 10^6	0.367	DCF in the effective zone [67]
10	Novikov pinion, m = 3.15 *mm*	NC	2174	2.30 10^6	0.845	DCF in the effective zone [116, 160]
11				*6.30* 10^6		Fracture [116, 160]

Note. In positions: 1-6 – flanks are ground after CHT; 7-11 – without finishing; 7 – durability is determined according to the average non-failure operation time; 10 – the actual hardness of specimens; 11 – distribution of hardness is statistic for the whole batch of specimens; (C – carburizing, NC- nitro carburizing).

Calculations of NG were carried out according to the initial geometry without the account of run-in. The taken coefficients are: $K_2=0.9$ (pos. 10), $K_2=1$ (pos. 11), $K_3=1$, $K_4=0.98$, $K_5=1$, $K_6=1$. Definition of initial geometry, force and other parameters was carried out: for involute gearing – according to [43], for NG – according to techniques, described in this monograph.

Analysis of the performed calculations results and their comparison with experimental data allows the making of the following conclusions:

1) The proposed model in most cases gives the satisfactory quantitative evaluation of DCS through the whole thickness of the diffusion layer. For hypoeutectoid structures (segments of the sublayer, adjacent to the core), the results approach

and for $H_{HV} < HV\,400$ coincide with the obtained ones according to the Mises criterion.

2) Among alternative models, the best correspondence is for models relating the allowable stresses with the hardness according to the Brinell scale. However, for the constant coefficient of proportionality, a systematic overestimation of DCS characteristics in the hard layer and their underestimation in the conversion zone takes place.
3) The technique [43] allows to determine satisfactorily the degree of maximum effective stresses action, but it does not provide the possibility of DCF occurrence in other zones of the layer. Moreover, the low accuracy of predicting the hardness distribution through the layer thickness and, in our opinion, the overestimated level of allowable stresses make inevitable application of big values of safety factors, which reached $1.25...1.47$ in performed calculations in spite of actually presented DCF.
4) The load-bearing capacity of the conversion zone is negatively influenced by the presence of the second "risky area", characterized by local reduction of design safety factors. In such cases, DCF can occur at greater depths (up to $z = 5\,b_H$), where the acting stresses are insignificant, and $S_{HK} > 1$ (Fig. 29). At the same time, when such a zone is absent and the monotony of increase (even insignificant quantitatively) of S_{HK} is kept, the relative strength of the layer is rather high. In experiments with the roller CA-120 (Fig. 28*b*), the maximum coefficient of proportionality *c* has been obtained. Within the limits of existing models, this situation can't be explained and, obviously, has not been investigated. That is why, when designing gearing, the parameters of the hardened layer must eliminate the appearance of the second "risky area".
5) Qualitative models demonstrate rather clearly the location of "risky areas" and can be applied at the initial stage of calculations in order to decrease their bulk.
6) Increasing the hardness of the hardened layer above $HV\,(650...700)$ for carburizing and $HV\,(700...750)$ for nitro carburizing practically does not lead to strength increasing, and further increase can give negative results because of the growing material embrittlement.
7) According to the DCS criterion, NG with the basic rack profile of RGU type and according to the standard GOST 30224-96 endure the increased specific contact loads, compared with involute one. For $\sigma_{k\,max} = 1900..2200 MPa$ and $N_{EK} = 10^6 ...10^7$ a single case of DCF (Fig. 31*a*), was registered, probably caused by the defect of the layer ("dip" of the hardness at the depth $0.3\ mm$, which could lead to the appearance of the second "risky area" – see Fig. 31*b*) and the presence of the large-needled martensite in microstructure [113]. In all other cases (more than 10), no DCF was revealed and the cause of failure was the fracture of teeth. This proves the correctness of increasing the allowable depth contact stresses for a theoretical point contact by more than 30% with respect to the established ones for the linear contact.

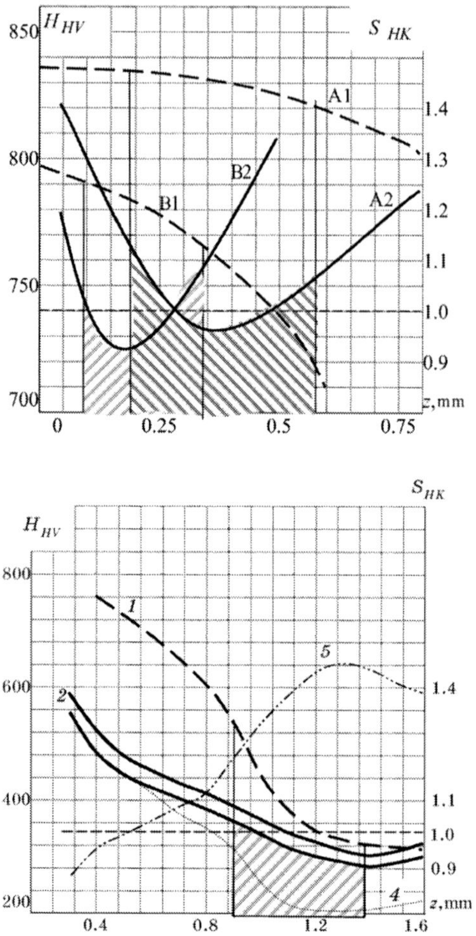

Figure 28. Characteristics of DCS for carburized rollers: *a*) – failures in the effective zone (A – roller CC-60; B – roller CD-30); *b*) – failure in the conversion zone (roller CA-120). Areas of incipient cracks are shaded.

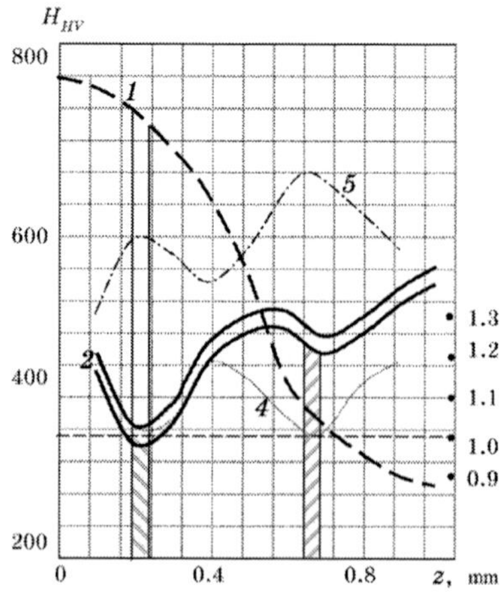

Figure 29. Characteristics of DCS, characterizing the breakage at two levels of carburized rollers CB-30.

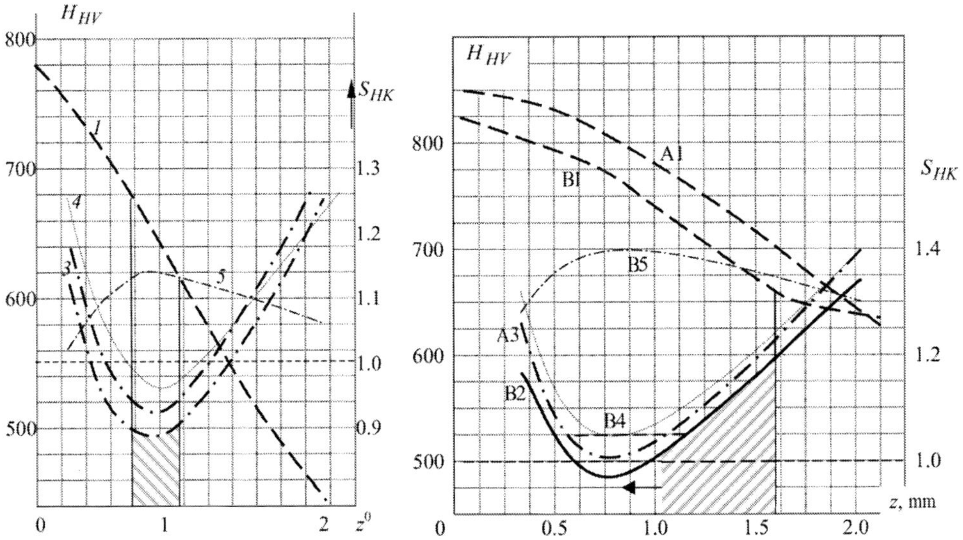

Figure 30. Characteristics of DCS for involute gearing: *a)* – pinion *m = 4.0 mm*; *b)* – pinion *m = 6.5 mm*; (A – carburizing, B – nitro carburizing).

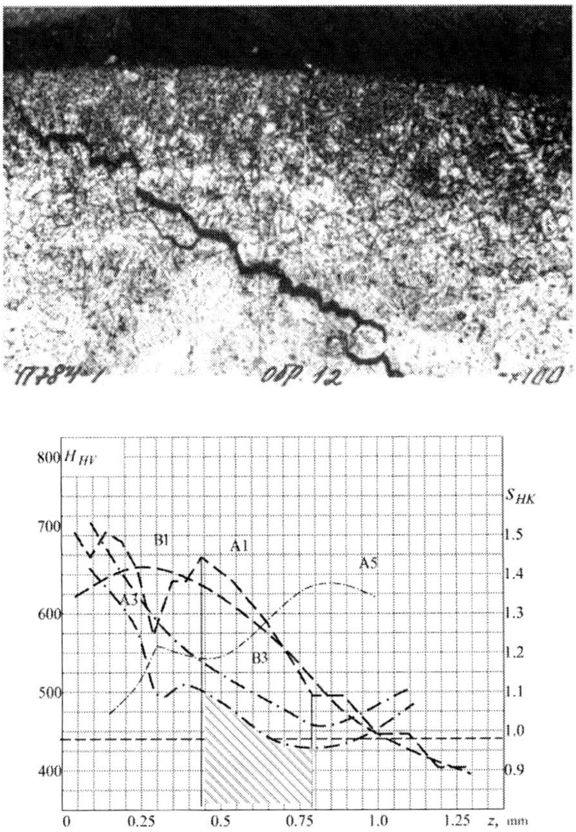

Figure 31. Novikov gearing: a) – lateral (transverse) microsection (pos. N10, Table 21); b) – characteristics of DCS (A – pos. N10, B – pos. N11 – Table 21).

12.5. Hardness Distribution within the Thickness of the Diffusion Layer

Since mechanical properties of steel are related in practice with its hardness as its simplest controlled characteristic, then the hardness distribution through the thickness of the diffusion layer for CHT is one of the main characteristics of the process quality.

Fig. 32 shows the typical hardness distribution for carburizing.

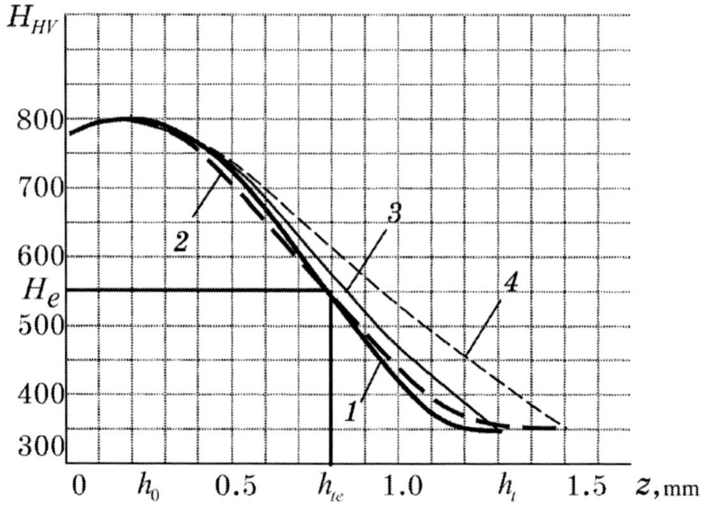

Figure 32. Characteristic view of diagrams of hardness distribution for carburizing: 1, 2 – according to the proposed relation (12.11) for $h_t = 1.3\ mm$ and $1.5\ mm$, correspondingly; 3, 4 – the same according to the relation (12.10).

At the stage of research development and design works, in the absence of experimental data, it is desirable to have either analytical or approximating relations of hardness distribution $H = f(z)$, where z is the design depth, with application of the main controlled parameters of the layer. These parameters are: H_0 – maximum hardness of the surface zone, H_k – hardness of the core, h_{te} – effective thickness of the layer – the distance between the surface and depth with the assigned level of hardness H_e (it is called effective). Mostly it is taken for carburizing $H_e = HV550$, for nitro carburizing $H_e = HV610$. Directly in the near-surface zone, the hardness H_n due to the inner oxidation, the presence of structures of non-martensite type etc., is lower than the maximum one, which is at the depth $h_0 = 0.15...0.30$ mm. The load-carrying capacity of the hardened layer is assumed to be defined exactly by this effective thickness. The total thickness of the layer h_t is, as a rule, not controlled because of its "blur". In [43], the relation is proposed for the current value H^z of the hardness with respect to the depth of the carburized and nitro carburized layer:

$$H^z = H_0[(H_0/H_k - 1)(z/h_t)^2 + 1]^{-1}, \qquad z \le h_t. \tag{12.10}$$

Fig. 32 shows the non-conformity of (12.10) in the area of sublayer – the segment, adjacent to the core, due to the great influence of the obstinate parameter h_t. Besides, $\frac{dH^z}{dz} \neq 0$ for $z = h_t$, and as a result of it, the hardness design values can significantly exceed the actual ones. This complicates the prediction of the appearance conditions of DCF sites in this zone. At the greater part of the interval, the approximating function must be low-sensitive to variations of h_t (within the limits of natural scatter and accuracy of measurements), it must pass through the checkpoint (h_{te}, H_e) and have $\frac{dH^z}{dz} = 0$ for $z = h_t$. According to the results of experimental data analysis, we obtained the relation for carburized layers as an approximating one [160, 161]:

$$H^z = (H_0 - H_k)\left[\frac{h_t - z}{h_t - h_0} exp(\frac{z - h_0}{h_t - h_0})\right]^B + H_k, \ z \leq h_t, \tag{12.11}$$

$$B = ln\left(\frac{H_0 - H_k}{H_e - H_k}\right) \cdot \left(ln\frac{h_t - h_0}{h_t - h_{te}} - \frac{h_{te} - h_0}{h_t - h_0}\right)^{-1}.$$

The diagram (12.11), shown in Fig. 32, demonstrates, that H^z, unlike (12.10), hardly depends on variations of h_t, reaching *0.2... 0.3 mm*.

The relation (12.10) is applied in reference literature, as it was noted above, for carburizing and nitro carburizing. However, the analysis of experimental data shows a number of differences. The decrease of the hardness H_n directly in the near-surface zone with respect to the maximum hardness H_0 is greater for nitro carburizing than for carburizing, and the area of the sublayer is more narrowed. That is why it is proposed to apply the following relation, differing from (12.11), for nitro carburized layers:

$$H^z = \frac{(H_n - H_0)(h_0 - z)}{h_0 - h_n} + H_0 \quad \text{for} \quad 0 < z \leq h_0, \tag{12.12}$$

$$H^z = (H_k - H_0)\left[\frac{z - h_0}{h_t} exp(1 - \frac{z - h_0}{h_t})\right]^B + H_0 \quad \text{for } h_0 < z \leq h_t,$$

$$B = ln\left(\frac{H_e - H_0}{H_k - H_0}\right) \cdot \left[ln\left(\frac{h_{te}}{h_t} exp\frac{h_t - h_{te}}{h_t}\right)\right]^{-1}.$$

The recommended values of H_e are for carburizing ≈ *HV 550*, for nitro carburizing ≈ *HV 610*. The attempts to introduce a new level of H_e, in most cases, worsened the correspondence of design values to experimental data.

Stochastic calculations, performed to check-up the proposed relations (12.11), (12.12), showed the decreased *1.5...2.5* times (compared with the results according to (12.10)) root-mean-square and absolute deviations of the design hardness from experimental values.

Therefore, the proposed relations allow to predict the hardness distribution through the diffusion layer depending on its main regulated parameters and types of CHT and can be applied when designing gearing, subjected to corresponding types of hardening.

12.6. Recommendations on the Choice of the Diffusion Layer Parameters

The proposed model of DCS evaluation and obtained relations of hardness distribution enable to specify a number of regulated parameters of the diffusion layer, namely – surface hardness, core hardness and effective thickness of the layer [114, 160].

It is recommended in existing regulations for involute gearing [43]: $H_0 = HRC\ (58...62)$, $H_k = HRC\ (30...45)$, $h_{te} = (0.18...0.27)m$ for carburizing and $(0.13...0.2)m$ for nitro carburizing. Obviously, these recommendations are averaged. Thus, the thickness of the layer is regulated according to the module, though (in other equal conditions) the level of contact stresses is not determined by the module, but by the center distance.

The effective thickness of the layer must be minimum, providing the required level of strength. For this purpose, the necessary hardness of the layer should be determined at the depth $z = (0.8...1.0)b_H$, that is, a little deeper than the level of maximum criterial stresses action, since in this very zone the intensity of reducing the material strength most often surpasses the intensity of reducing σ_e. Then, assigning H_0 and H_k, according to (12.11) or (12.12), the required effective thickness is determined. Here, the total thickness of the layer on the basis of experimental data processing can be roughly taken:

$$h_t = h_{te}(H_0 - H_k)/(H_0 - H_e) \text{ - for carburizing,}$$
$$h_t = h_{te} + (1.45 - \gamma \cdot H_k) \text{ - for nitro carburizing,} \qquad (12.13)$$

where $\gamma = 0.002 - 0.0025$ for $H_k \geq HV350$ and $\gamma = 0.003$ for $H_k < HV350$.

The maximum hardness of the surface zone is not the limiting factor from DCS point of view, however, it influences as one of parameters of hardness distribution through the thickness of the hardened layer. If the value h_{te} is fixed, the increase of H_0 leads to reduction of the relative thickness of the layer, that is, to weakening of the conversion zone. That is why it is desirable to have minimum hardness H_0, providing the required level of the load-bearing capacity with respect to alternative types of failure: pitting, wear, fracture, scuffing resistance, etc.

If H_0 is assigned in a rather narrow range, caused by the inevitable scatter of material properties and manufacturing factors, then the variation of $H_k = HRC\ (30...45)$ [43] is allowable only in the case when the core hardness does not essentially influence functional properties of the gearing. It is no mere chance that the load-bearing capacity of carburized and nitro carburized layers is estimated in standards ISO, DIN, etc. only with respect to their effective zone. However, the total operation experience of heavy-loaded gearing hardened by CHT shows the positive influence of increasing the core hardness on the level of load-bearing capacity of nitro-carburized parts [66, 210].

The comparative analysis of the stressed state of carburized and nitro carburized diffusion layers for variation of H_k within the range *HV (300...440)* showed considerably different influence of these variations depending on the type of CHT, though in any case the reduction of H_k requires a definite increase of the hardened layer thickness.

For reasonably chosen parameters H_0, H_k, h_{te} ,the potentially dangerous area (with minimum safety factor) is within the limits of the effective zone, and the load-bearing capacity of the gear is specified by the strength of the layer itself. However, if we reduce H_k, for instance, from *HV 350* to *HV 300* (that is, keeping the standard recommendation), remaining invariable the value of the effective thickness, then the total growth of the hardened layer thickness (due to the sublayer) for nitro carburizing, related to it, can become insufficient, without compensating the hardness reduction. As the result, another potentially dangerous zone appears, which in a number of cases becomes leading (Fig. 33*b*). Moreover, the presence of several "risky areas" considerably reduces the durability of the gear, even if the safety factor remains more than 1 in these areas [114].

A somewhat different situation is observed for carburizing (Fig. 33*a*). The reduction of H_k (for invariable h_{te}) leads to more considerable growth of the sublayer, complicating the appearance of the second "risky area". In our opinion, this is one of the reasons of the carburized layers decreased sensitivity to the level of H_k.

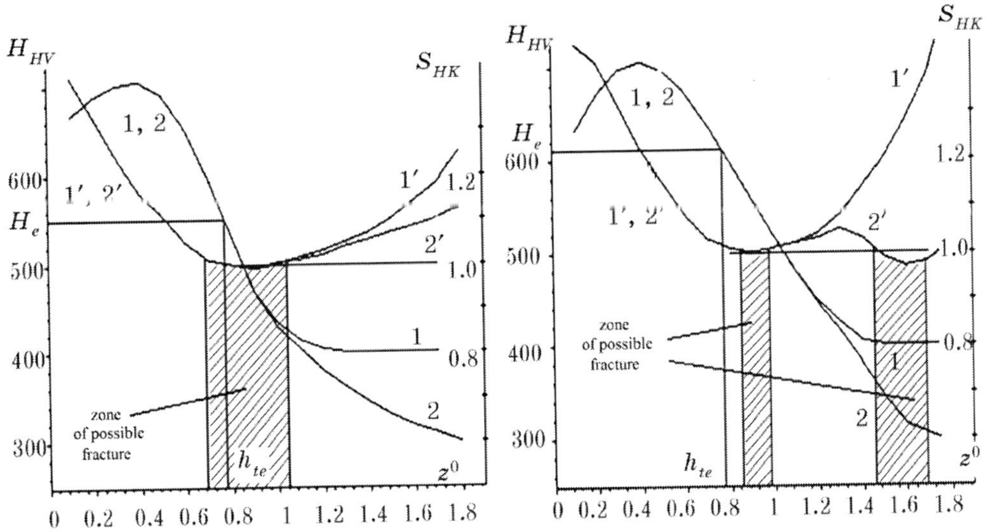

Figure 33. Influence of the core hardness on the load-bearing capacity of the carburized (*a*) and nitro carburized (*b*) diffusion layer: 1, 1' – distribution of hardness and strength safety factor S_{HK} for H_k = *HV 400*;2, 2' – distribution of hardness and strength safety factor S_{HK} for H_k = *HV 300.*

At the same time, the analysis of the stressed state of hardened layers shows sufficiently high strength and stability of sublayer zones within the hardness range *HV (400...500)*. Failure sites appear either higher – in the area of maximum equivalent stresses action, or lower – in the area of transition of the sublayer to the core. The latter does not allow the usage of high strength characteristics of the diffusion layer, forcing either to reduce the level of load, or to increase the effective thickness of the layer.

Therefore, the influence of the core hardness is essentially different for carburized and nitro carburized parts, and the value H_k must be regulated depending on the type of CHT. And if it is allowable to keep the existing recommendation H_k =*HV (300...440)* for carburizing, then for nitro carburized gears it is recommended H_k > *HV (350...380)*.

12.7 Nitrided Gearing

In previous paragraphs, only two types of CHT have been considered, applied for achieving the maximum load-bearing capacity of gearing – carburizing and nitro carburizing. Along with the noted types, nitriding is widely used in gearing production.

This type of CHT has a number of undoubted advantages. High hardness and wear resistance of the surface zone are achieved when nitriding. The temperature of nitriding does not exceed *560...600* degrees, which does not cause phase recrystallization. Thermal deformations and surface roughness are such, that the final machining is not provided, even for important and high-speed gearing. Power intensity of the process is lower, than for carburizing and nitro carburizing [251].

However, application of nitriding for high-stressed gearing is limited by a comparatively small (under *0.6 mm*) thickness of the hardened layer and core hardness, which is caused by the preliminary heat treatment of steel (normalizing, refining) and does not exceed in practice *HB (300...320)*. The operation experience of nitrided, in particular, involute, gearing shows considerably greater scatter of results on load-bearing capacity compared with carburizing and nitro carburizing.

Letus consider the conditions of providing DCS in involute and NG, hardened by gas nitriding, to the effective thickness of *0.35mm*. The thickness of the layer under $H_e = HV400$ will be considered effective [15]. Hardness distribution is taken according to [31] for nitrided layers:

$$H^z = 0.8(H_0 - H_k)(z/h_t)^2 - 1.8(H_0 - H_k)(z/h_t)^{1.1} + H_0. \qquad (12.14)$$

Characteristics of DCS, defined according to the proposed design model, are given in Fig. 34 for typical examples. Their analysis allows for the making of the following conclusions:

1) The intensity of hardness, and, therefore, strength variation through the thickness of the diffusion layer is considerably higher for nitriding than for carburizing and nitro carburizing.
2) In involute gearing, the area of maximum criterial stresses action is comparable with the thickness of the hardened layer, and, along with the intensive variation of the hardness in the layer, it defines the increased sensitivity of nitrided involute gearing to overloads and variations of the layer thickness. Reduction of the core hardness can essentially decrease the level of contact-fatigue durability due to appearance of the second “risky area”, without essential influence on the effective zone itself.
3) In NG, which should be considered as run-in after nitriding of one of the pair elements (usually pinion), the depth of maximum criterial stresses occurrence

can significantly (*2...3* times) exceed technically achievable thickness of the diffusion layer, that is why in this type of gearing, the level of contact-fatigue durability will be determined only by the strength of the core. Appearance of the additional concentrator as the second "risky area" is practically excluded. Therefore, in order to achieve the maximum depth contact strength, it is required to perform preliminary high-quality heat treatment, allowing to obtain high strength characteristics of the tooth core.

4) Compared with the calculation of carburized and nitro carburized gearing, the calculation of nitrided gearing is simplified, since in this case we have: $Z_{LK} = 1, \chi = 1, K_2 = 1, K_3 = 1, K_5 = 1, K_6 = 0.95$.

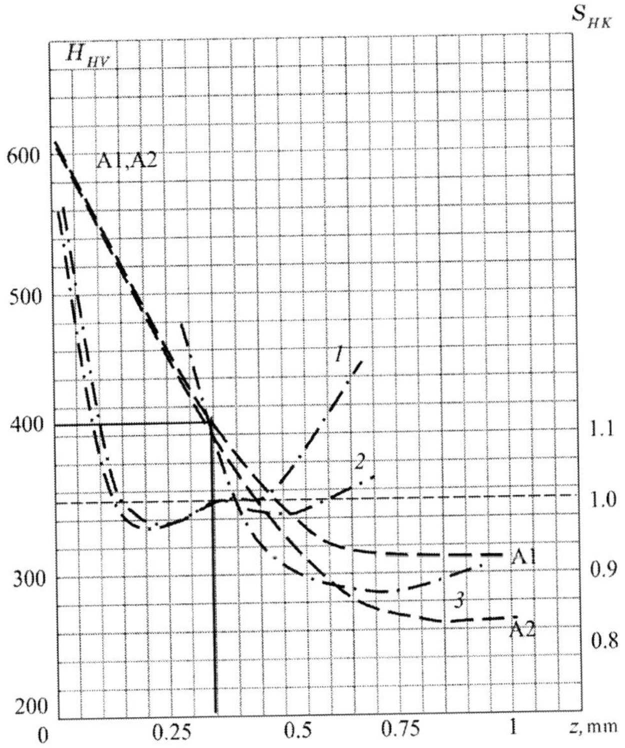

Figure 34. Characteristics of the depth contact strength for nitriding: A1, A2 – hardness distribution through the layer thickness for $H_{HV} = HV\ 300$ and $H_{HV} = HV\ 250$; Design safety factors: 1 – involute pinions, $m = 2.25\ mm$, $\sigma_{k\,max} = 1600 MPa$, $N_{EK} = 10^7$, $b_H \approx 0.20\ mm$, $H_k = HV\ 300$; 2 – the same for $H_k = HV\ 250$; 3 – Novikov gearing (basic rack profile RGU-5): $m = 3.15\ mm$, $\sigma_{k\,max} = 1516 MPa$, $N_{EK} = 10^7$, $b_H \approx 1.013\ mm$, $H_k = HV\ 300$.

12.8 Methodical Recommendations on Calculation of Novikov Gearing, Hardened by CHT, for Prevention of DCF

The purpose of the calculation is to determine parameters of the hardened layer and type of CHT, providing the required level of load-bearing capacity of the gearing (design analysis), or to define the allowable load-bearing capacity of the gearing for given parameters of the diffusion layer and type of CHT (checking analysis) according to the condition of the depth contact failures absence. As the result of calculation, the safety factors according to DCS are determined (as the ratio of the allowable contact stresses to the acting ones) in different depth layers and areas of minimum contact endurance "risky areas" are revealed.

It was noted above, that for nitriding (usually for the pinion only), the NG should be calculated as run-in (for $Z_{LK} = 1$). As for carburizing and nitro carburizing, both elements of the gear pair are subjected to these types of CHT, the run-in is less intensive and, as it was observed in experiments [196], the depth cracks appear more often at a small number of cycles of operating time. That is why, in these cases, the calculation should be performed for not run-in surfaces ($Z_{LK} = 1.125$).

Initial data for calculation: $\sigma_{k\,max}$ – maximum surface contact stress; b_H – minor semi-axis of the contact ellipse; $\overline{\beta}$ – ellipticity coefficient; grade of steel; required type of CHT; final (after CHT) machining of contacting surfaces; availability of automated control of the CHT process; H_0, H_k - maximum hardness of the tooth flank and core hardness correspondingly;

When calculating steel gears, the modulus of elasticity is usually taken as $E = 2.15 \cdot 10^5$ MPa and Poisson's ratio as $\nu = 0.3$.

The effective hardness is taken as (see the paragraph 12.5): $H_e = HV\,550$ for carburizing and $H_e = HV\,610$ for nitro carburizing.

Values of $\sigma_{k\,max} b_H, \overline{\beta}$ for unrun tooth surfaces are determined on the basis of (11.22), (11.24), (11.25), the design normal load F_{Hn} is defined according to (10.2) by the design torque T_H (10.5), assuming that $K_\lambda = 1, K_f = 1$ (10.7).

For run-in surfaces, one can write:

$$\sigma_{k\,max} = t_e \left[F_{Hn} E^2 / (\overset{*}{\rho}_\beta m)^2 \right]^{1/3}, \tag{12.15}$$

where t_e is calculated by complete elliptical integrals K, F of the module e (e is the eccentricity of the contact ellipse) [96]:

$$t_e = e \left[e / (K - F)^2 \right]^{1/3} / 3\overline{\beta}. \tag{12.16}$$

The parameter t_e depends on $C_{\alpha\beta} = \rho_\alpha / \rho_\beta$, and it can be approximated within the range $C_{\alpha\beta} = 0.01 \ldots 0.3$ $(\overline{\beta} \leq 0.45)$ with the accuracy up to $\pm\,2\%$ by the power dependence

$$t_e = 0.345C_{\alpha\beta}^{-0.44}. \quad (12.17)$$

Then, using (11.2), (12.15), (12.17), (10.2), we obtain for $E = 2.15 \cdot 10^5$ MPa and $\nu = 0.3$:

$$\sigma_{k\,max} = 8626\left[T_{H2}/(z_2 \cos\alpha_k)\right]^{0.606} m^{-1.82} (Z_l K_l l^*)^{-0.821} (\rho_\beta^*)^{-0.394} \quad (12.18)$$

Comparing (11.5) and (12.15) with account of (11.6) and (12.17), we get

$$t_e / K_e = \sigma_{k\,max} / \sigma_{He} = 4.72C_{\alpha\beta}^{0.13}, \quad (12.19)$$

hence, after corresponding substitutions and transformations:

$$C_{\alpha\beta} = 3.574\left[z_2 \cos\alpha_k /(T_{H2}\rho_\beta^*)\right]^{0.623} (Z_l K_l l^* m)^{1.87}. \quad (12.20)$$

It follows from (12.20), that the less the transmitted load (limited by bending or surface contact endurance of gearing), the more the value of $C_{\alpha\beta}$ (and, therefore, the value of the reduced curvature profile radius ρ_α for the invariable value of the longitudinal radius ρ_β) increases after the surface run-in compared with $C_{\alpha\beta}$ (and ρ_α) for the initial geometry. Thus, the increase of $C_{\alpha\beta}$ during the run-in process for example, for the thermally improved gearing is more considerable than for gearing with surface hardened teeth, which is rather logical.

Letus give the design analysis procedure.

1) Assigning the components of (12.7): correction coefficients K_2, K_4, K_5, K_6 and durability factor Z_{LK}.
2) Definition of the necessary thickness of the hardened layer.
   a) The conditional design compression stress is taken:
   $$(\sigma_{k\,max})_{con} = \sigma_{k\,max}/(K_2 \cdot K_4 \cdot K_5 \cdot K_6 \cdot Z_{LK}).$$
   b) According to $(\sigma_{k\,max})_{con}$ and diagrams 3 and 4 (Fig. 27) the required hardness of the layer H_{kr} is determined at the depth of maximum criterial stresses action.
   c) χ_{kr} is the parameter of the material plasticity at the depth of maximum criterial stresses action, it is defined according to (12.6) and data from the paragraph 2.2.
   d) The depth z_{kr}^0 of maximum criterial stresses occurrence is determined from Table 19 depending on the contact ellipticity and plasticity parameter according to the paragraph 2.3.
   e) The effective thickness h_{te} of the hardened layer is determined according to auxiliary diagrams in Fig. 35 depending on the type of CHT and the

depth of occurrence of maximum criterial stresses according to the paragraph 2.4.

f) The total thickness h_t of the hardened layer is determined according to (12.13).

g) The thickness of the layer, corresponding to the hardness H_0: if there is final machining of contact surfaces $h_0 = 0$, without it – $h_0 = (0.10...0.30)\ mm$ (C) and $(0.20...0.40)\ mm$ (NC).

3) The hardness distribution through the hardened layer is determined according to (12.11) or (12.12). The calculation is performed in steps for different depths with the step $(0.1...0.25)\,b_H$.

4) Preliminary definition of potentially dangerous areas. According to the auxiliary Table 22, the values σ_i/H_{HV} for various depths z^0 are determined within the hardened layer (with possible plotting of the auxiliary diagram). Areas, characterized by values $(\sigma_i/H_{HV})_{max}$ or closeness to zero of the derivative of the ratio σ_i/H_{HV} are fixed. Further calculation is carried out only for these areas, the depths of their occurrence is designated as z'.

5) Acting standardized effective stresses. According to (12.1): $\sigma_e^0 = \sigma_i \cdot \chi - (1-\chi) \cdot R$, where the values of the coefficient R are given in Table 22. Values of the plasticity parameter χ for the fixed points, determined according to the item 4, are calculated according to (12.6).

6) Allowable stresses σ_{HKPe}^0, regulated with respect to $\sigma_{k\,max}$, are determined according to (12.7).

7) The safety factor according to DCS: $S_{HK} = \sigma_{HKPe}^0 / \sigma_e^0$.

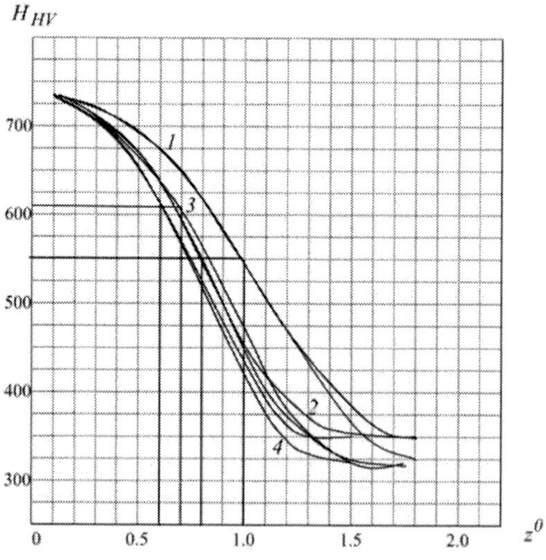

Figure 35. Approximate diagram of hardness distribution, used in design calculations of DCS: 1 – carburizing, $h_{te} = 1.0\,z^0$; 2 – carburizing, $h_{te} = 0.8\,z^0$; 3 – nitro carburizing, $h_{te} = 0.7\,z^0$; 4 – nitro carburizing, $h_{te} = 0.6\,z^0$.

If the obtained results are unsatisfactory, parameters of the hardened layer are corrected, and the calculation (items 3-7) is repeated.

Calculation of nitrided gearing.

Since the high load-bearing capacity of the effective layer itself is not achieved in NG, due to great absolute depths of maximum criterial stresses occurrence, the reasonable type of CHT in many cases can be nitriding. For run-in nitrided gearing, the level of DCS is determined by the core hardness, the coefficients from (12.7) are accepted as follows: $K_1 = 2.7$ (maximum possible value), $K_2 = 1, K_3 = 1, K_4 = 1.04, K_5 = 1, K_6 = 0.95, Z_{LK} = 1.$

Since $H_k < HV400$, then $\chi_{kr} = 1, \sigma_e = \sigma_i$, and $z^0_{kr} = 0.6...0.7$ (depending on $\overline{\beta}$). Then, according to the condition $\sigma_{i\,max} \le \sigma_{HKPe}$ (for widespread cases $\overline{\beta} < 0.30$), we obtain the condition of the depth contact strength:

$$\sigma_{k\,max} \le (3.95...4.0) H_k. \quad (H_k \text{ - according to Vickers}). \tag{12.21}$$

Table 22. Components of the effective stress depending on the depth of its occurrence and ellipticity of contact.

k_e / z_0	0		0.10		0.20		0.30		0.50	
	σ_i	$R \cdot 10^2$	σ_i	$R \cdot 10^2$	σ_i	$R \cdot 10^2$	σ_i	$R \cdot 10^2$	σ_i	$R \cdot 10^2$
0.10	0.400	7.32	0.364	6.73	0.349	6.48	0.342	6.45	0.344	6.83
0.30	0.477	10.33	0.475	10.90	0.484	11.27	0.497	11.32	0.522	11.39
0.50	0.540	10.86	0.550	10.08	0.546	9.66	0.578	9.36	0.600	8.90
0.70	0.558	8.28	0.572	7.38	0.586	6.75	0.597	6.32	0.607	5.67
0.90	0.546	6.01	0.562	4.96	0.573	4.32	0.579	3.87	0.576	3.23
1.10	0.519	4.32	0.535	3.22	0.543	2.60	0.544	2.17	0.527	1.65
1.30	0.486	3.13	0.501	2.02	0.506	1.44	0.502	1.07	0.473	–
1.50	0.453	2.31	0.476	1.21	0.468	–	0.459	–	0.422	–
1.75	0.413	1.62	0.428	–	0.423	–	0.410	–	0.363	–
2.00	0.378	–	0.388	–	0.383	–	0.364	–	0.313	–
2.25	0.347	–	0.351	–						
2.50	0.319	–	-	-						
3.00	0.274	–	-	-						

$$R = -\sigma_1 \cdot 0.75^{\left|1-\frac{\sigma_1+\sigma_2+\sigma_3}{\sigma_i}\right|}$$

Chapter 13

BENDING STRENGTH ANALYSIS OF CYLINDRICAL NOVIKOV GEARING

13.1. TOOTH BENDING ENDURANCE ANALYSIS

Bending strength is of a special importance for Novikov gearing.

As it was already mentioned, the first attempts to implement NG with hard tooth flanks failed exactly because of their insufficient bending strength, which was the consequence of the unreasonable approach to the design of the basic rack profile and the gearing as a whole.

Theoretical point, and local character of load application to the tooth, peculiar to NG, creates a higher level of the tooth SSS, than the theoretical linear contacting, when the load is uniformly distributed along the tooth. This seems to be the reason of conclusion that linear contact is advantageous.

However, the given argumentations are valid for the idealized scheme only.

In reality, even insignificant errors of gears manufacturing , their assembly (such as misalignment and obliquity of axes, errors of the axial pitch, direction of the tooth line etc), and deformations of parts under load during operation inevitably lead to transformation of the linear contact into the local, moreover, to the edge one, which immediately converts the theoretical advantage of the linear contact into its drawback, causing the increased concentration of the load and stresses along the width of toothing and the increased dynamic load. Such situation in the involute gearing is partially improved by the application of teeth, barrel-shaped along the length, by profile flanking, etc.

Due to the local character of teeth contacting and good run-in of their flanks, NG is less sensitive to various misalignments in the engagement (angular and tangential errors), than the involute one, and the considerable tooth compliance creates more favorable dynamic conditions and enables more uniform distribution of the load and stresses in the multiple-tooth contact. This probably explains the fact that the correctly designed NG with hard unground teeth possesses not only higher contact, but also bending endurance, than the involute analogs (see Chapter 8).

Another important advantage of NG should be emphasized. With an increase of the module for the constant center distance, not only fracture strength is increased in NG, but also the contact one (including depth strength), creating in a number of cases the effective additional reserve of increasing its total load-bearing capacity.

Letus pass on to the brief description of the state-of-the-art on tooth bending endurance analysis of NG. Two main approaches should be pointed out here.

The first approach implies the creation of semi-empirical engineering formulas based on experimental results. This approach is developed in works of LMI [140], where the results of nature tests have been applied, and also in works of ZAFEA [172] and UZPI [8], where the results of experiments on definition of stresses in three-dimensional models have been used.

The obvious fact does not need to be proved that calculation techniques, based on the first approach, do not possess versatility, and they agree well with practical results only within the range of the performed experiments.

The second approach, possessing greater versatility, is based on the application of one or another conditional scheme of tooth loading, having analytical or numerical solution. Without the detailed analysis of proposed schemes and hypotheses by different authors, where certain assumptions are certainly present, we note that the scheme, according to [194], is the most widespread, where the solution is applied of the plane problem according to the method of cylindrical sections [164] combined with the solution of the cantilever plate stressed state under the action of the concentrated force [261].

The new impulse to improve bending analysis of NG teeth was given by the appearance of numerical methods (finite-element method – FEM, boundary integral approach – BIA, variation methods, etc.) and the application of modern computers in the practice of investigations and calculations.

During many years, simultaneous activity was carried out on both specification of the two-dimensional tooth SSS of the complex shape [4, 5, 6, 168, 171, 181] and on solution of the three-dimensional task by numerical methods [11, 248, 250].

The level of SSS of NG teeth is determined by solving the three-dimensional contact problem with unknown in advance contact area for helical teeth with finite height and length subjected to the multiple tooth contact and taking into account manufacturer errors. Such a serious problem has not been solved completely until the present, though its separate points are described in publications.

Solution of the SSS three-dimensional task, satisfactory for practical implementation, is described in [250], where the method of BIA is applied, the longitudinal dimension of the contact area being related to bending shift deformations and calculated iteratively by means of a computer. Solution of this problem allowed to calculate the volume factor Y_v of the tooth shape, which was subjected to the unit concentrated force, and the coefficient Y_a, depending on the longitudinal contact area extension along the tooth length.

It should be noted that the structure of the formula for bending stresses determination is obtained similarly both in [194] and in [250], however, in [250], the definition of the tooth shape coefficient is better scientifically based:

$$\sigma_F = 2000 K_F T Y_v Y_a / (m^3 z), \tag{13.1}$$

where K_F is the load coefficient, described above (10.8).

The volume factor Y_v of the tooth shape depends on basic rack profile geometry, reduced number of teeth z_v and coefficient x^* of the basic rack profile shift at cutting, it can be taken according to diagrams (Fig. 36-40) [176] or calculated by the polynomial:

$$Y_v = \sum_{0}^{3} A_{0q}(x^* + 1)^q + z_v^{-1} \sum_{0}^{4} A_{1q}(x^* + 1)^q + z_v^{-2} \sum_{0}^{4} A_{2q}(x^* + 1)^q, \tag{13.2}$$

where coefficients of the polynomial are assigned according to the Table 23.

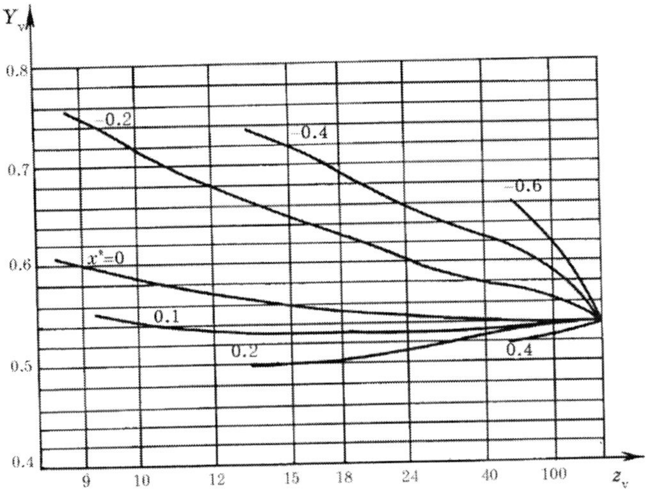

Figure 36. Diagram for definition of volume coefficient Y_v of the gear tooth shape of the gearing with the basic rack profile DLZ-0.7-0.15.

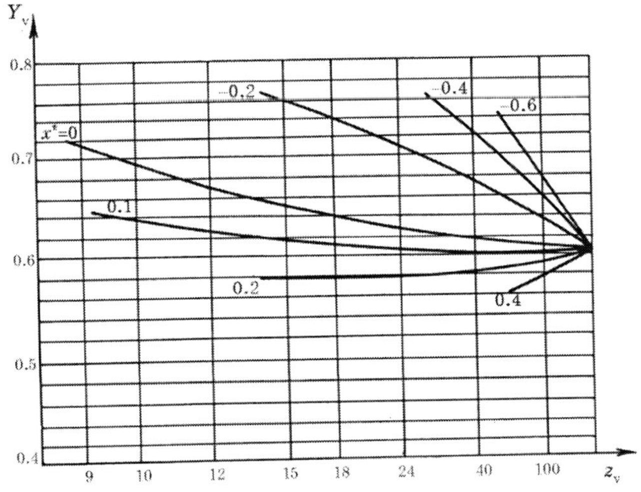

Figure 37. Diagram for definition of volume coefficient Y_v of the gear tooth shape of the gearing with the basic rack profile DLZ-1.0-0.15.

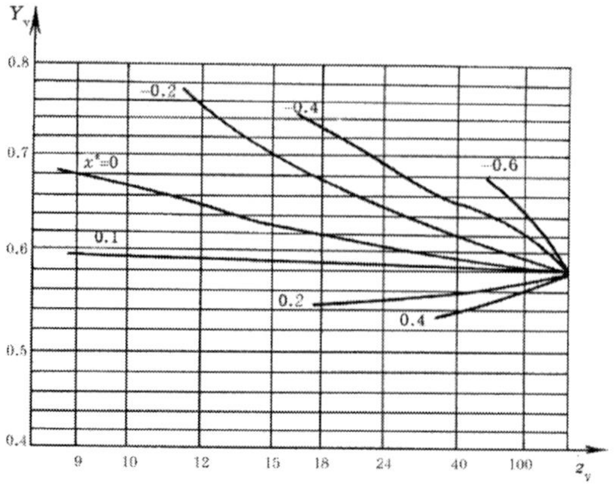

Figure 38. Diagram for definition of volume coefficient Y_v of the gear tooth shape of the gearing with the basic rack profile according to the standard GOST 15023-76.

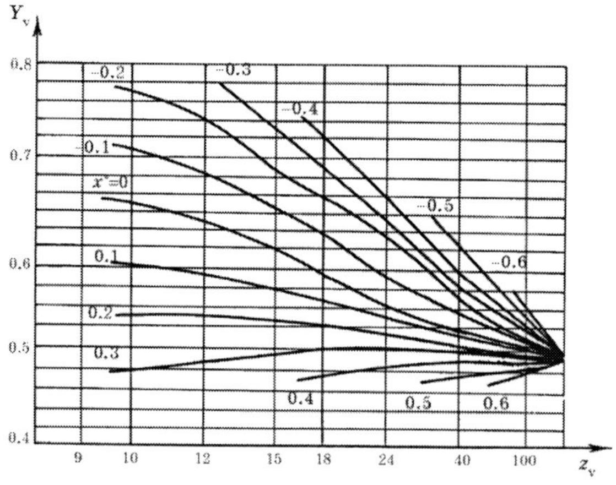

Figure 39. Diagram for definition of volume coefficient Y_v of the gear tooth shape of the gearing with the basic rack profile RGU-5.

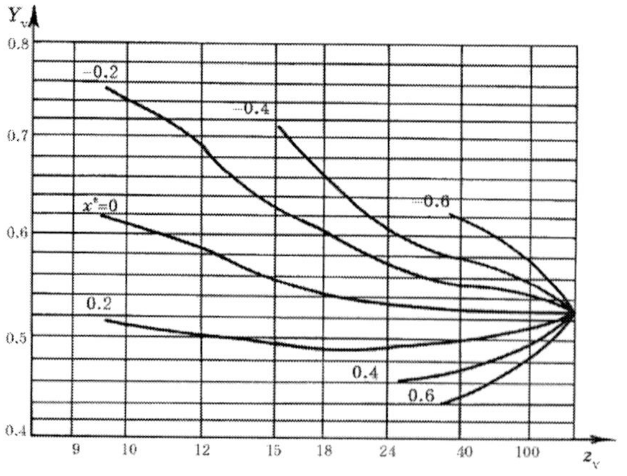

Figure 40. Diagram for definition of volume coefficient Y_v of the gear tooth shape of the gearing with the basic rack profile according to the standard GOST 30224-96.

Table 23. Coefficients of polynomial (13.2).

A	Basic rack profile		
	RGU-5	According to the standard GOST 15023-76	According to the standard GOST 30224-96
A_{00}	0.58044	0.48597	0.51602
A_{01}	-0.30638	0.42992	0.13316
A_{02}	0.29535	-0.53941	-0.18046
A_{03}	-0.08988	0.20211	0.062714
A_{10}	2.3856	34.193	14.986
A_{11}	10.357	-119.92	-55.553
A_{12}	-15.978	158.46	78.540
A_{13}	4.7696	-89.326	-49.084
A_{14}	0.40658	17.117	10.692
A_{20}	-17.231	-375.08	-153.44
A_{21}	192.76	1325.3	809.64
A_{22}	-453.30	-1477.3	-1265.1
A_{23}	390.26	545.64	791.82
A_{24}	-115.43	-14.991	-170.58

The coefficient Y_a is approximated by the relation:

$$Y_a = 1 - t_2 (a_F^*)^2 (t_3 + t_4 / z_v) / [1 + t_1 a_F^* + t_2 (a_F^*)^2], \tag{13.3}$$

where $a_F^* = a_H / m$, a_H - see (11.24), $t_1 - t_4$ are taken according to the Table 24.

Table 24. Values of coefficients t_1-t_4 from (13.3).

Basic rack profile	t_1	t_2	t_3	t_4
DLZ-0.7-0.15	0.24	0.14	1.15	-2.70
DLZ-1.0-0.15	0.21	0.11	1.15	-2.70
According to the standard GOST 15023-76	0.14	0.07	1.19	-3.42
RGU-5, according to the standard GOST 30224-96	0.13	0.07	1.13	-2.34

As it is seen from (13.3), that the coefficient Y_a depends on the load, the relation between the stress σ_F and the load is non-linear: σ_F is varying slower than T_F.

Taking into account the strength condition

$$\sigma_F \le \sigma_{FP}, \tag{13.4}$$

letus write the expression for the allowable stress σ_{FP} from [43], calculated separately for the pinion and the gear:

$$\sigma_{FP} = \sigma^0_{F\,lim\,b} Y_z Y_N Y_A Y_\delta Y_X / S_F. \tag{13.5}$$

Here the designations are [43]:

$\sigma^0_{F\,lim\,b}$ is the bending endurance limit of teeth, corresponding to the number of loading cycles, equal to $4 \cdot 10^6$, depending on the grade of steel, type and parameters of heat treatment [43] and taken as a reference point for determining the allowable stresses;

Y_z is the coefficient, accounting for the methods of the gear preforming, equal to *1* for forged pieces and stampings, *0.9* for rolled metal and *0.8* for cast workpieces;

Y_N is the durability factor, equal to

$$Y_N = \sqrt[q_F]{4 \cdot 10^6 / N_F}, \tag{13.6}$$

but not less than *1*; for gears with homogeneous material structure $q_F = 6$, for gears with surface hardening of teeth $q_F = 9$; maximum values $Y_{N\,max} = 4$ for $q_F = 6$, $Y_{N\,max} = 2.5$ for $q_F = 9$. (When applying the method of equivalent cycles, N_{FE} is substituted instead of N_F);

Y_A is the coefficient, accounting for the influence of the two-sided load application, calculated according to the relation

$$Y_A = 1 - \gamma_A [T'_F Y_N / (T_F Y'_N)], \tag{13.7}$$

where T'_F, Y'_N are the load and the durability factor correspondingly for calculation of the opposite side of the tooth; γ_A is the coefficient, equal to *0.35* for gears made of annealed, normalized and thermally improved steel, *0.25* for gears with surface hardening (except for nitrided) and *0.1* for nitrided gears; Y_δ is the reference coefficient, equal to

$$Y_\delta = 1.082 - 0.172 \log m; \tag{13.8}$$

Y_X is the coefficient, taking into account the gear dimensions:

$$Y_X = 1.05 - 0.000125d; \tag{13.9}$$

S_F is the safety factor, see Chapter 14 devoted to its choice.

13.2. Bending Strength Analysis under the Action of Maximum Load

The strength condition for this analysis is expressed as

$$\sigma_{F max} \leq \sigma_{FPmax}. \quad (13.10)$$

Knowing the stresses σ_F (13.1) under the load T_F, one can determine the stresses $\sigma_{F max}$ under the load $T_{F max}$ according to the approximating conversion formula with account of the noted above non-linear relation:

$$\sigma_{F max} / \sigma_F = (T_{F max} / T_F)^{0.8547}. \quad (13.11)$$

The allowable stress can be determined according to the formula [43]:

$$\sigma_{FP max} = \sigma^0_{FSt} Y_X / S_F, \quad (13.12)$$

where σ^0_{FSt} is a certain basic value of the tooth extreme bending stress under maximum load, depending on the steel chemical composition and the type of its heat treatment [43]; Y_X - according to (13.9); S_F is described in Chapter 14.

Chapter 14

CHOICE OF SAFETY FACTORS IN CONTACT AND BENDING ENDURANCE ANALYSIS OF CYLINDRICAL NOVIKOV GEARING

The checking analysis of a gearing in general, is known to be based on satisfaction of the strength condition

$$\sigma \le \sigma_P \tag{14.1}$$

where σ is the calculated acting stress, σ_P - the allowable stress.

In order to determine the acting stress, some or other schemes of the tooth stressed state are developed, taking into account the maximum possible number of factors influencing the working capacity of the gearing. In advanced research works, the calculation of σ is developed, gradually approximating the adequate model of real tooth operation conditions.

Allowable stress is assigned either directly on the basis of experiments, or through a certain limiting stress σ_{lim} and a safety factor S:

$$\sigma_P = \sigma_{lim}/S. \tag{14.2}$$

The second approach is applied in standard analyses of involute gearing [43], where median experimental values of endurance limits were taken as σ_{lim}, the values having a rather conditional nature, that is reflected in [43] by stipulations about the possibility to change the indicated values, proved by rig and nature tests. Essentially, designers of the standard [43] consider the noted values σ_{lim} as a convenient reference point for determination of allowable stresses.

As for the safety factor S, let's consider it in detail.

For many years in widely spread gearing analysis methods, the safety factor was interpreted as a parameter, the introduction of which to the gearing analysis secures it from the negative influence of possible material properties instability, and the recommended value of which due to this reason is connected with the type of the material heat treatment, method of the preforming, degree of the gearing significance, etc. [140, 172, 176]. Here, the necessity of organic connection of the coefficient S with the left part of the strength condition (14.1), that is, with the degree of adequacy of the design model to the real load-bearing capacity of the gearing, is concealed. Such an interpretation of the coefficient S leads to the fact that, in practical analysis, its value and the whole right part of the inequality (14.1) is kept constant regardless of some or other corrections of the design model. This, in turn, leads to an absurd conclusion of real tooth strength dependence on the applied method of its analysis. For NG, analysis methods are known [140, 176], in which values of the right part of the inequality (14.1) completely coincide with those for involute gearing, though calculation of acting stresses (left part of inequality) for the noted types of engagement is principally different. The obtained mismatch of both parts of the inequality (14.1) is sometimes compensated by the introduction of unreasoned and even unexplained correction parameters [140].

Of course, instability of the material properties (even when the structure corresponds to specified technical conditions) can't help influencing the variation of the gearing load-bearing capacity, however, this and similar factors, connected with material, must be taken into account when assigning values of σ_{lim} with a given reliability, that is successfully done in [43]. In the same work, in our opinion, the concept of a safety factor was given for the first time, as a parameter, integrally taking into account the approximate nature of the analysis method. We think that this concept contains the functional meaning of the coefficient S, consisting in bringing to conformity the left part of the strength condition (14.1), depending on the analysis method, with the right part, depending on the material properties. That is why the correction of the design scheme must be accompanied with the correction of the coefficient S (at constant σ_{lim}), which tends to be reduced as this scheme is developed.

Thus, the coefficient S (which, in our opinion, due to its function could be more suitably named "a balancing coefficient") is closely connected with the method and structure of the acting stresses analysis.

Determination of the coefficient S is an important task, as its insufficient value can cause gearing premature failure, sometimes with serious consequences, and its overestimated value – to unreasonably increased overall dimensions and weight of the drive. In both cases there are considerable economical losses.

To validate the particular values of the coefficient S, it is necessary to have at least:

a) the design model, based on modern achievements of the elasticity theory, in which the factors influencing the gearing operation should be taken into account as fully as possible;
b) sufficient number of reliable results (desirably, which are statistically processed) of the gearing tests in the wide range of its parameters variation.

When fulfilling these necessary conditions, it is possible to expect receiving in the first approximation (not excluding the possibility of further correction) values of safety factors S, invariant to wide-range variation of geometrical, manufacturing, stiffness and other gearing parameters, influencing the level of the tooth stressed state.

Having as a test result the experimental allowable load T_P^E and substituting it into the design formula of the left part of (14.1), we will obtain the design stress, equal to the limiting one, i.e. $\sigma = \sigma_P$, after that, taking a certain value σ_{lim} as a basis, according to (14.2) we determine $S = \sigma_{lim}/\sigma = \sigma_{lim}/\sigma_P$.

The design formula for determination of the acting bending tooth stress σ_F, obtained theoretically, is validated above (13.1), where the load factor K_F is described in Chapter 10 (10.8).

Values of conditional limiting stresses σ_{Flim} with all accompanying correction coefficients, recommended in [43], were chosen as a reference point for the determination of allowable stresses.

An experimental base was represented by test results, completely corresponding to the technical conditions of NG with various basic rack profiles, geometry, degree of accuracy, hardness of tooth flanks, carried out during 30 years in SFU together with Izhevsk "OOO "Reduktor" [22, 129] and LMI [3].

Table 25 contains the main parameters of the tested gearing.

Table 25. Parameters of gear pairs compared according to experimental and design data.

Parameter	Gearing N					
	1	2	3	4	5	6
Basic rack profile	RGU-5			According to [39]	According to [45]	
a_w, *mm*	160	250		160	130.2	150
m, *mm*	3.15	5	6.3	3.15	4	
z_1	32	18	13	32	29	12
z_2	65	75	62	65	30	59
b_w, *mm*	45	63		45	36	47
β, *deg.*	17.284	21.565	20.223	17.284	25	18.797
Type of heat treatment	Nitro carburizing	Martempering		Nitro carburizing	Martempering	
H_{HB1}	570	302	270	570	269	241
H_{HB2}	570	269	230	570	255	207
Tested by	SFU (RSU) – "OOO "Reduktor"				LMI	

According to [40], the gearing had the degree of accuracy within the range *8...9*.

During the gearing bending endurance tests, the number of stress alternation cycles in all experiments greatly exceeded the basic number of cycles of the driven gear. A maximum number of tests were carried out with the gearing *N1*, that made it possible to carry out correlation analysis of the results and to obtain the allowable load T_{FP}^E with the probability of teeth non-failure $P = 0.995$ (see Chapter 8). For other types of gearing, sufficiently reliable, allowable or limiting fracture loads $T_{F\,lim}^E$ were obtained.

Results of determination of the fracture safety factor S_F in accordance with (13.1) and (14.2), are shown in the Table 26 [111].

On the basis of data of Table 26, it is possible to recommend in the first approximation $S_F = 1.2$ (gearing *N1*) for surface hardened (for example, nitro carburized) teeth, and $S_F = 1.1$ (gearing *N2* and *N5*) for thermally improved and normalized teeth.

Table 26. Design and experimental data for bending endurance of the compared gear pairs.

Parameter	Gearing *N*				
	1	2	3	4	5
$T^E_{F\,lim\,2}$, $N{\cdot}m$	-	-	10000	3050	-
T^E_{FP2}, $N{\cdot}m$	2181 ($P = 0.995$)	6250	-	-	797
$\sigma_{F\,lim}$, MPa [43]	1033	453	380	1033	451
σ_F, MPa (13.1)	870	409	375	1057	406
S_F (14.2)	1.187	1.108	-	-	1.111

Noted in [3], description of gearing *N6* test allows to suppose that its allowable load is at the level of *1500...1600 N·m*. Substituting in (13.1), the value σ_{FP}, determined according to [43] for $S_F = 1.1$, and solving the equation relative to the load, we will obtain $T_{FP} = 1473$ *N·m*, that indicates that the value $S_F = 1.1$ for the given gearing is, in any case, not underrated.

For the gearing *N3* and *N4,* only limiting loads could be obtained, substitution of its values into (13.1) gives, as it is seen from the Table 26, stress values, close to the limiting ones $\sigma_{F\,lim}$, recommended by [43] that, in a certain extent, confirms the successful choice of the latter as a reference point for finding σ_{FP}.

Let's proceed to the contact endurance analysis.

Here the design relationship (11.8), as it was noted, is based on the determination of effective contact stresses, reduced to normal ones, and taking into account the run-in of tooth surfaces in their contact, depending on the surface hardness.

Several remarks should be made concerning contact endurance tests of the gearing.

Because of the known duration of these tests (especially with hardened tooth surfaces), many researchers carry them out under a limited number of cycles, that requires commenting the test results by indicating the number of cycles, at which they were obtained. On the other hand, the criteria of the gearing failure according to surface pitting often causes uncertain evaluations, since it is unstable and "blurred". That is why, to our mind, it is more reasonable to fix in tests those loads, at which rather clean tooth surfaces are obtained without visible contact defects in repeated tests. In this case, corresponding loads may be characterized qualitatively, determining a certain threshold, below which the probability of the actual allowable loads existence is rather low.From such a standpoint, the contact endurance test results were analyzed, the tests were carried out by SFU (former RSU) together with "OOO "Reduktor" and LMI [3] and described in Chapter 8.

On the basis of the stated reasons, it is possible to make a conclusion that the determination of the safety factor S_H according to (11.8) and (14.2), as it was similarly done for determining S_F, is not reasonable. It is more logical to act in another way – having specified the allowable stress σ_{HP}, to find the allowable design load T_{HP} by means of (11.8), and to compare it with the experimental one T_{HP}^E. The results of such an analysis are shown in Table 27 [111], where the stress values, recommended in [43] (see Chapter 11), taking into account the durability factor Z_N for the given number of cycles and for $S_H = 1$, are taken as σ_{HP}, i.e., it is accepted that $\sigma_{HP} = \sigma_{Hlim}$.

Table 27. Design and experimental data for contact endurance of the compared gear pairs.

Parameter	Gearing N					
	1	2	3	4	5	6**)
T_{HP2}^E, N·m	Not less than 3047	6250*)	Not less than 5008	Not less than 3555	702	1005
Number of cycles of the pinion N_{H1}, mln	52.5	52.5	80.5	51.0	52.0	107.0
σ_{HP}, MPa [43]	1551	635	506	1558	588	455
T_{HP2}, N·m (11.8)	3560	4935	5745	3440	690	981

*) weak pitting appeared;
**) increased radial error of gears manufacturing.

As it is seen from Table 27 for $S_H = 1$, the allowable design load T_{HP} either practically coincides with the experimental one T_{HP}^E (gearing *N5* and *N6*), or exceeds it a little (gearing *N1* and *N3*), which is qualitatively allowable due to the reasons stated above. For the gearing *N2* and *N4*, the design load appeared to be even less than the experimental one. It shows that it is possible, at least, in the first approximation and not excluding further possible correction, to apply in analyses, regardless of the tooth surfaces hardness, $\sigma_{HP} = \sigma_{Hlim}$ ($S_H = 1$), where σ_{Hlim} should be determined according to recommendations of [43]. If we had applied $S_H > 1$ in the given analysis, we would have obtained allowable design loads, which vary from the experimental ones more considerably, than at $S_H = 1$.

The conclusion, which can be made on the basis of the stated information, is as follows:

When applying the design formulas (11.8) and (13.1), it is possible to accept values, recommended in [43] as limiting stresses σ_{lim} both on bending and contact endurance, and to accept, in the first approximation, as safety factors:

- in bending endurance analysis $S_F = 1.2$ and 1.1 for surface hardened (besides nitrided) and thermally improved teeth (nitrided also), respectively;

- in contact endurance analysis $S_H = 1$ regardless of the hardness of contacting tooth flanks.

Chapter 15

RECOMMENDATIONS ON ANALYSIS OF CYLINDRICAL NOVIKOV GEARING UNDER VARIABLE LOADS

It is known that many mechanisms (general-purpose gearboxes, crane gearboxes, gearboxes for oil pumping units and others) work under variable loads, the level and duration of which are changed according to rather complicated and different laws. It is rather difficult to imitate the real situation of the drive loading during its analysis completely. As the result, some typical loading modes are developed for different drives, which simulate the operation modes with a known approximation. Thus, for example, for speed gearboxes of cars and machine-tools four loading types are accepted, which are subdued to different statistical laws [151]. For general-purpose gearboxes, operating in the intermittent mode, step loading diagrams are accepted, for which light-duty, medium-duty and heavy-duty modes are marked out [187].

The main technique for taking into account variable loads, changing in time gradually or according to a step diagram, when analyzing the Novikov gearing, can be accepted similarly to the stated in [43]. Here, depending on the analysis purpose, one of three methods can be used: method of equivalent cycles, method of equivalent moments or method of equivalent stresses.

When carrying out the checking analysis for determining the equivalent number of stress alternation cycles, it is convenient to apply the first of the noted methods, finding N_E as

$$N_E = \mu N, \tag{15.1}$$

where the coefficient μ allows to take into account the character of the loading cyclogram.

Usually the maximum load, T_{max} according to the cyclogram ($T_H = T_{max}$, $T_F = T_{max}$), is applied as a design load in variable loading modes; the time of its action, expressed in cycles N_{Tmax}, is not less than $0.03\ N_{Hlim}$ in contact endurance analysis, and not less than $5 \cdot 10^4$ in bending endurance analysis [43].

Then, for the step cyclogram, we have

$$\mu = \sum_i (T_i / T_{max})^{K_q} N_i / N_{T\,max}, \tag{15.2}$$

for the gradual character of the cyclogram

$$\mu = \int_{T\,min}^{T\,max} (T_i / T_{max})^{K_q} \, d(N_i / N_{T\,max}), \tag{15.3}$$

where the "i"-th load T_i of the cyclogram corresponds to the number N_i of cycles of its action.

Assuming that the oblique part of the fatigue diagram, representing the relationship between stresses and cycles of the stress alternation, for NG is the same as for involute gearing (see power exponents q_H, q_F – Chapters 11, 13), the power exponent K_q in formulas (15.2) and (15.3) should be determined in NG by other, than for involute gearing, relationship between loads and stresses (11.8), (13.11):

- i
  n contact endurance analysis $K_{Hq} = 0.687q_H$; (15.4)
- i
  n bending endurance analysis $K_{Fq} = 0.8547q_F$. (15.5)

The example is given in Fig. 41, where a typical step cyclogram for general-purpose gearboxes of heavy-duty mode [187] is shown. According to this cyclogram, assuming that $q_H = 6$, $q_F = 9$, we have: $\mu_H = 1 + 0.75^{4.12} \cdot 2.5 + 0.2^{4.12} \cdot 1.5 = 1.766$; $\mu_F = 1 + 0.75^{7.69} \cdot 2.5 + 0.2^{7.69} \cdot 1.5 = 1.274$.

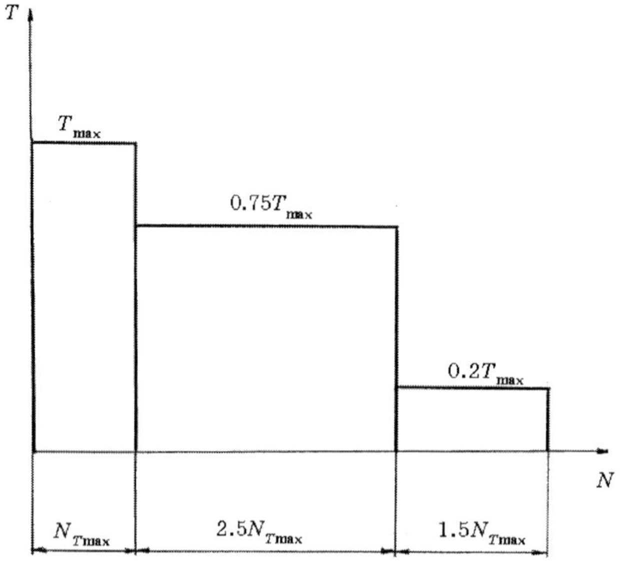

Figure 41. Typical step-type cyclogram for heavy-duty operation mode of general-purpose gearboxes.

The equivalent number of cycles in contact endurance analysis is $N_{HE} = 1.766 N_{T\,max}$ (for the design load $T_H = T_{max}$), and in bending endurance analysis is $N_{FE} = 1.274 N_{T\,max}$ (for the design load $T_F = T_{max}$).

Chapter 16

DESIGN ANALYSIS OF CYLINDRICAL NOVIKOV GEARING

The design analysis is carried out to determine preliminary (approximate) geometrical parameters of the gearing and doesn't replace the checking analyses, in which these parameters must be corrected and finally accepted.

The design analysis is based on the principle of assigning the maximum possible module, because in NG, this allows to achieve the increase of both bending and contact strength of teeth [129, 239], under other equal conditions. On this basis, it is advisable to have the total tooth number z_Σ of the pair for high-hardened teeth under *60...62*, for thermally improved teeth – under *83...85*, as it is shown in practice.

Initial data for the design analysis:

- accepted basic rack profile of teeth;
- cyclograms of torques and speeds;
- material, type of heat treatment, hardness of surfaces and core of the pinion and gear teeth;
- operation mode (constant, reverse, etc.);
- design time of the pair operation;
- gear ratio u of the pair;
- helix angle β;
- axial overlap ratio ε_β.

Note.

Taking into account the discrete nature of NG, application of some ranges of ε_β is not reasonable. That is why it is recommended to assign ε_β within the limits *1.1...1.2* or *1.6...1.7*, and in the extreme case (mainly, for the gearing with low tooth hardness) – *2.1...2.2*.

The analysis is carried out as follows.

1) The total number of teeth z_Σ of the pair is assigned according to recommendations, stated above.
2) The number of pinion teeth is determined:

$$z_1 = z_\Sigma/(u+1) \geq 9. \tag{16.1}$$

The restriction $z_{min} = 9$ is related to the possibility of teeth undercut for very small numbers of teeth, and also with inexpediency of application in certain cases of cut-in pinion-shafts.

3) ρ^*_β is determined according to (2.20).
4) l^* is determined according to (3.18).
5) Z_l is determined according to (11.9).
6) $K_H = 1, K_l = 1$ are accepted preliminary.
7) The minimum allowable stress σ_{HP} for the pinion and gear is determined according to [43], taking into account the stated above recommendations on assigning the safety factor S_H (Chapter 14).
8) The gearing module is calculated according to the condition of the contact endurance:

$$m_H = (Z_H/\sigma_{HP})^{0.485}[T/(z\cos\alpha_k)]^{0.333}(\rho^*_\beta)^{-0.151}(Z_l l^*)^{-0.515}. \tag{16.2}$$

9) Y_v is found from diagrams in Fig. 36-40, preliminary taking $x^* = 0$.
10) Assigning $a^*_F = 2$, Y_a is determined according to (13.3).
11) $K_F = 1$ is preliminary accepted.
12) The allowable stress σ_{FP} is determined according to [43], taking into account the stated above recommendations on assigning the safety factor S_F (Chapter 14).
13) The ratios Y_v/σ_{FP} at the pinion and gear are determined; when they differ greatly, shift factors x^*_1, x^*_2 are chosen so that the noted ratios for the pinion and gear become close to each other, tending to the strength equality of their teeth, by changing the values Y_{v1} and Y_{v2} (determined from diagrams in Fig. 36-40). If it is not possible to make the noted ratios Y_v/σ_{FP} equal, then the maximum obtained value for the pinion or gear is chosen for further analysis.
14) The gearing module is determined according to the condition of bending endurance:

$$m_F = 12.6[TY_vY_a/(\sigma_{FP}z)]^{1/3}, \tag{16.3}$$

15) The maximum value is chosen from two values m_H (16.2) and m_F (16.3)

$$m = max\{m_H, m_F\}. \tag{16.4}$$

Note 1. If in the designed gearing $\varepsilon_\beta \geq 1.6$, then the module value, found according to (16.4), should be reduced by $(10...15)\%$ before the checking analysis.

Note 2. For the herring-bone gearing, designed according to parameters of the half width of the herring-bone, the value of the module, found according to (16.4), should be reduced by *(15...20)%*.

The value of the module, obtained in the design analysis, is rounded according to [37].

16) The operating width of toothing b_w of the gear pair is determined:

$$b_w = \pi m \varepsilon_\beta / \sin\beta . \tag{16.5}$$

17) It is accepted for the gear that $b_2 = b_w$, or a little greater for design reasons, and for the pinion $b_1 = b_w + \Delta b$, where Δb is accepted for design reasons.

After completion of the design analysis, the full geometrical and checking strength analyses are carried out using the techniques, stated above, with simultaneous correction of gearing parameters, if it is necessary.

Chapter 17

BEVEL NOVIKOV GEARING

In this chapter, only bevel Novikov gearing with circular teeth will be considered, since gearing with another longitudinal tooth shape (for example, helical, cycloidal [191] and others) is practically not applied.

17.1. GENERAL INFORMATION ON INVESTIGATIONS AND APPLICATION

The first prototype bevel gears with a new type of engagement were produced under the supervision of M.L. Novikov. Since 1956, theoretical and experimental works have been carried out by E.G. Ginzburg, who studied methods of synthesis of the conjugate point engagement of the gearing with crossed axes, based on the applyication of the concept of a generating pair and machine-tool engagement; he developed systems of geometrical analysis and tooth cutting by means of the existing equipment. The scheme of the cutting system was based on determination of relationships between gears dimensions and radius of the cutter head, at which tooth undercut by auxiliary cutting edges of cutters near the internal face and possibility of "ridges" appearance at roots near the external face were excluded [35].

The first scientists who investigated the bevel NG in Russia, were E.G. Roslivker, I.S. Dreizen [192], S.I. Minchenko [153], Ya.S. Davydov [55], A.M. Badaev [7]. They solved the main problems of the manufacturing technique, designed and analyses as applied for the gearing with one line of action, noted advantages of the gearing with respect to contact strength and revealed disadvantages, which occurred because of sensitivity to manufacture and assembly errors, causing the shift of the contact area along the tooth height (radial errors).

Further development of bevel NG is connected with the application of the gearing with two lines of action. The works [17, 57, 147, 154, 170, 172, 220] represent investigations of the gearing and technique of geometrical and strength analyses, contain information about manufacture, accuracy and control of gears. The geometry problems are quite thoroughly developed in [201-203], where a generalized theory of circular helical surfaces is proposed, on the basis of which a gearing with given quality characteristics can be designed and methods of contact strength and bending strength analyses can be developed.

Experimental investigation of bevel NG, improvement of its design and manufacturing technique, implementation into the national economy are represented in works [93, 94, 138].

An original system of tooth cutting (a system “Riga”) is proposed by K.K. Paulinsh [174, 175], in which a special cutter head with circular shape cutters is applied. In this cutter head, the generating radius of the cutters can be changed continuously and in a rather wide range, that is convenient for experimental works. The cutters are simple in manufacturing, but their small number in the cutter head prevents this system from application in large-batch and mass production. At the same time, it is possible to combine the tooth shaping and ensure longitudinal modification by application of the four-sided cutter head.

Solution of an important problem of rising the tooth bending strength in Novikov gears by increasing the module at the expense of decreasing the axial overlap ratio along one line to less than unit is described in works [27, 28].

Among the research works devoted to the problems of bevel Novikov gears accuracy, the works [48, 143, 246] can be pointed out, directed to the generation of mathematical model of the real engagement, allowing to study a set of problems, which are necessary to be solved during the development of accuracy standards and tolerances for manufacturing and assembly technique.

Along with revealing the design peculiarities of bevel NG, methods of its manufacturing are analyzed. Generation according to the method of plane generating gear [172] is mainly applied. Both special casings of cutter heads and standard ones [36] are used here, equipped with circular [175], sharpened [167] or profiled [236] cutters. An important advantage of the latter is their capability to retain dimensions of the generating surface when resharpening. However, there are certain technological problems during their manufacturing because of the surface complexity, since axial sections of generating and face flank surfaces of cutters do not coincide [76].

Methods of cutters manufacture, analysis of the grinding wheel profile [77, 144, 150], cutters control are developed.

Thus, a considerable number of works on investigating the bevel NG have been carried out so far. The problems devoted to geometry of engagement are thoroughly studied, but there are still certain difficulties of assigning the degrees of accuracy for tool and gear manufacturing, and also of carrying out the strength analysis.

It should be noted, that the existing research and practical realization of the gearing are directed to the generation of the conjugate NG. In this case for TLA gearing, the theory of engagement is applied, developed for the gearing with one line of action without taking into account the correlation of contacting along two lines of action at the presence of manufacturing and assembly errors. But, compared with the quasi-involute bevel gearing, the conjugate NG is more sensitive to radial errors, that requires its improvement in order to widen its industrial application.

At present, the bevel NG with the basic rack profile according to the State standard GOST 15023-76 and carburized teeth is batch- produced at the Ust-Katavsk Rail Carriage Building Plant. The application of NG allowed to design a single-stage gearbox for a drive of a tram KTM-5M with the gear ratio $u = 7.143$, to reduce the gearbox weight and decrease production intensity [93]. And these results could be reached in spite of the fact that, as it was already noted, application of the basic rack profile according to the State standard GOST 15023-76 with flank hardness more than *HB 320* is not rather effective because of the low fracture strength of teeth and small contact strength of the near-pitch-point area, which begins to operate at definite conditions, and high sensitivity to errors [233].

Bevel NG is successfully applied in gearboxes of coal cutter-loaders [201], in repair works of the trolleybus main gearboxes [28] and other cars [94, 137, 138]. The system "Riga" is implemented at Liepaysk Machine-building Plant [175], at Kungur Machine-building PO and others.

In order to increase the working capacity of bevel NG by reducing its sensitivity to manufacturing errors, on the basis of the stated below methods of analysis and synthesis of approximate gearing, taking into account the presence of several lines of action, their mutual influence on the gearing kinematics and conditions of the tooth contact, the gearing should be applied with localized contact along the tooth length (with a longitudinal modification of surfaces), also the advanced methods of the strength analysis (see the paragraph 17.7) should be applied.

17.2. Schemes of Tooth Generation

Novikov bevel gear teeth are produced mainly by tooth cutting generation method. During the tooth cutting, the process is simulated of machine-tool engagement of the cut gear with a planar imaginary generating gear, tooth flank of which (generating surface) is formed by rotating cutting edges of the cutter head. To obtain a theoretically correct conjugation of the pinion and gear teeth of TLA gearing, a rigid non-congruent generating pair [146] should be formed, i.e. to form two unmatched generating surfaces touching each other at two lines. Gear workpieces are mounted on the machine-tool so that the pitch cone of the cut gear and the pitch plane of the generating gear could touch each other and the pitch cone apex was superposed with the axis of the generating gear. When the described scheme of the conjugated tooth cutting is theoretically precise, the geometry of engagement is mainly determined by the technological method of machining (one-side or two-side) and parameters of the generating surface [201]. The longitudinal tooth shape entirely depends on the diameter of the cutter head.

As it is known, kinematically precise conjugate gears don't provide high working capacity of the power quasi-involute gearing [148] because of errors. It will be shown below, during the analysis of the kinematics of engagement, that conjugate bevel NG is also sensitive to some manufacture and assembly errors, which becomes apparent in unbalanced operation of the gearing along two lines of action [231], smoothness violation of rotation transmitting, the contact shift to the tooth edge and, as a consequence, in increasing the contact stresses, worsening the vibroacoustic characteristics. It is known that in order to adapt the teeth of quasi-involute bevel gears to operation under high variable loads, the purposive variation of the machine-tool settings is applied, localizing tooth contact along the tooth length and height [146, 204]. In NG, the contact localization along the tooth height is provided by the basic rack profile shape. To obtain the longitudinal localization, a certain increase of the cutter head generating diameter is required, which forms the concave tooth side, or its decrease when forming the convex tooth side, as compared with heads, cutting a rigid non-congruent generating pair. Here the value of the contact localization and changing the tool diameter are rigidly interrelated which leads to certain problems with tools in piece and small-batch production with large nomenclature of the cut gears. However, manufacturing problems regarding tools are compensated to some extent by the simplicity of the machine-tool settings calculation, as it will be shown later.

A more general modified scheme of cutting the bevel gear teeth (Fig. 42), based on the application of the machine-tool engagement with crossed rotation axes of the generating and the cut gear (machine-tool engagement of a hyperboloid gearing), has much more flexibility with respect to providing the required contact longitudinal localization and obtaining a reasonable longitudinal teeth narrowing without cutter heads change [235].

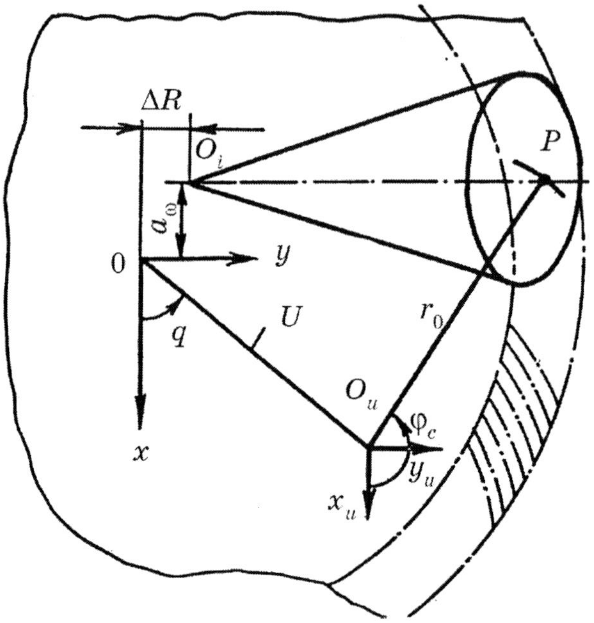

Figure 42. Modified scheme of bevel gear tooth-machining in projection on the plane, perpendicular to the axis of generating gear rotation.

Fig. 42 shows the scheme of a machine-tool engagement projected to the pitch plane of the generating gear, and Fig. 43 – its projection to a plane drawn through axes *1,2* of rotation of the cut pinion and gear. The pitch cones of the pinion and the gear touch at the generatrix $O_iP = R$, lying in the pitch plane *xOy* of the generating gear. The apexes of pitch cones O_i *(i* = 1,2*)* are shifted relative to the axis *3* of the generating gear in a direction of a common perpendicular to axes *1, 2, 3* of the generating and the cut gears to the value of a center distance a_w and along the generatrix of the pitch cone O_iP to a distance *ΔR*. Fig. 42 shows the positive shift of a_w, *ΔR*. Axes of the generating gear and generating surface are projected to points O, O_u, correspondingly. The fixed right coordinate systems *S (x, y, z)*, $S_1''(x_1'', y_1'', z_1'')$, $S_2''(x_2'', y_2'', z_2'')$ are applied. The coordinate axis *Oy* is drawn in the pitch plane parallel to the generatrix O_iP of the pitch cone, the coordinate axis *Oz* coincides with the axis *3*, the origins of the coordinate systems S_1'', S_2'' are placed to apexes O_i of pitch cones, axes z_1'', z_2'' are superposed with axes of rotation *1, 2*, and axes y_1'', y_2'' are drawn parallel to the plane *yOz*.

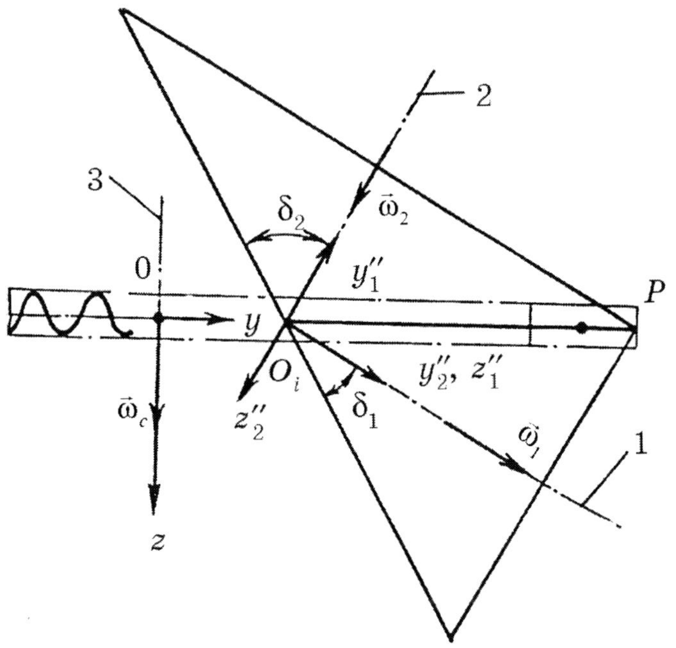

Figure 43. Modified scheme of bevel gear pair tooth-machining in projection on the plane, passing through the axes of gears rotation.

The rotation axis $O_u z_u$ of the generating surface is shifted with respect to the axis Oz for the value U of the radial setting. When shaping the tooth surface by a generating surface with the design radius r_0, a design (usually mean) normal helix angle β_n is provided at the design point P. The generating surface is shown in Fig. 44 within the system $S_u(x_u, y_u, z_u)$. The axis z_u coincides with the axis of the cutter head and is displaced with the machine-tool cradle during generation. At the same time, there is a change of a run-in factor – the angle q between the coordinate axis Ox and the straight line OO_u, passing through the axes of the generating gear and the pitch surface in the pitch plane, and axes x_u, y_u remain parallel to the axes x, y. The axial section of the generating surface of rotation is considered to be given in a parametric form:

$$\begin{cases} z_u = z_u(\vartheta); \\ \rho = \rho(\vartheta). \end{cases} \tag{17.1}$$

The generating surface, formed by rotation of the line (17.1) around the axis z_u, will be described by vector equation in the system S_u:

$$\bar{r}_u = \rho\cos\varphi_c \bar{i} + \rho\sin\varphi_c \bar{j} + z_u \bar{k} \tag{17.2}$$

Here ρ, φ_c, z_u - cylindrical coordinates, ϑ - parameter of an axial section of the generating surface.

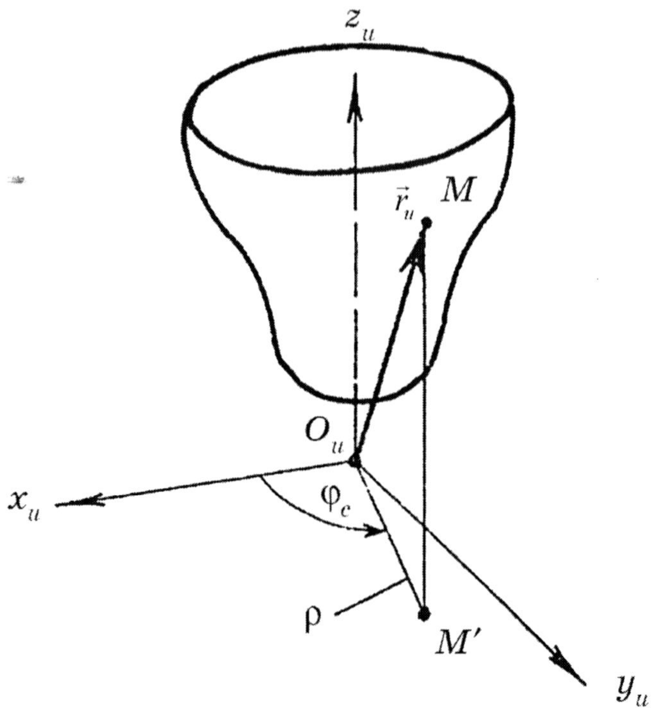

Figure 44. Generating surface for cutting the bevel gear.

In a particular case, for a circle with the radius ρ_0 , cylindrical coordinates and their derivatives are determined by relations: $\rho = R_0 + \rho_0 \cos\vartheta,\ z_u = z_0 + \rho_0 \sin\vartheta,\ \frac{d\rho}{d\vartheta} = -\rho_0 \sin\vartheta,\ \frac{dz_u}{d\vartheta} = \rho_0 \cos\vartheta$,

where R_0, z_0 are coordinates of the circle center.

By applying a kinematic method [146] of surface envelope determination, the equation of the machine-tool engagement is found as

$$F_c = \overline{V} \cdot \overline{n} = \frac{dz_u}{d\vartheta}\left[(1 - i_c \sin\delta)/(i_c \cos\delta)U \sin(q - \varphi_c) + (-1)^n z_u \cos\varphi_c + \right.$$
$$\left. + \tan\delta(\Delta R \cos\varphi_c + a_w \sin\varphi_c)\right] + (-1)^n \frac{d\rho}{d\vartheta}(\rho \cos\varphi_c + U \cos q + a_w) = 0 \qquad (17.3)$$

where $\overline{V}$ is the relative speed vector of the generating and the cut surfaces tangency points in the machine-tool engagement, $\overline{n}$ is the vector of the common normal to these surfaces, $i_c = \omega/\omega_c$ is the ratio of the cut and the generating gears angular speeds, δ is the pitch cone angle, n is the power exponent, accepted depending on the direction of the tooth line: for right gear $n=1$, for left gear $n=2$.

The vector of the common normal of the generating and the cut surfaces in the coordinate system S is determined by the expression $\overline{n} = \frac{dz_u}{d\vartheta}\cos\varphi_c \overline{i} + \frac{dz_u}{d\vartheta}\sin\varphi_c \overline{j} - \frac{d\rho}{d\vartheta}\overline{k}$.

Letus transform the equation (17.3) to the form $a_c^* \sin\varphi_c + b_c^* \cos\varphi_c = c_c^*$ and determine the parameter

$$\varphi_c = 2\,arctan\left\{\left[a_c^* \pm (a_c^{*2} + b_c^{*2} - c_c^{*2})^{0.5}\right]/(b_c^* + c_c^*)\right\} \tag{17.4}$$

where

$$a_c^* = \frac{dz_u}{d\vartheta}\left[a_w \tan\delta - U\cos q(1 - i_c \sin\delta)/i_c/\cos\delta\right];$$

$$b_c^* = (-1)^n \frac{d\rho}{d\vartheta}\rho + \frac{dz_u}{d\vartheta}\left[U \sin q(1 - i_c \sin\delta)/i_c/\cos\delta + (-1)^n z_u + \Delta R \tan\delta\right];$$

$$c_c^* = -(-1)^n \frac{d\rho}{d\vartheta}(U\cos q_c + a_w).$$

The extraneous root φ_c can be excluded by comparing the received roots of the equation with an approximate value $\varphi_c = arccos\dfrac{U\cos q + a_w}{z_u/\tan\vartheta - \rho}$. In some cases, when both root values of the equation (17.4) are close to each other, this equation has to be solved additionally, by reducing $sin\varphi_c$ and $cos\varphi_c$ to a cosine of the auxiliary angle.

Taking into account (17.4), the equation of the tooth flank of the cut gears will be obtained [146]:

$$\begin{cases} x_i = \cos\theta(a_w + \rho\cos\varphi_c + U\cos q) + \sin\theta\sin\delta \times \\ \times(\rho\sin\varphi_c + U\sin q - \Delta R - (-1)^n z_u \cot\delta); \\ y_i = \cos\theta\sin\delta(\rho\sin\varphi_c + U\sin q - \Delta R - (-1)^n z_u \cot\delta) - \\ -\sin\theta(a_w + \rho\cos\varphi_c + U\cos q); \\ z_i = (-1)^n \cos\delta(\rho\sin\varphi_c + U\sin q - \Delta R) + z_u \sin\delta, \end{cases} \tag{17.5}$$

where $\theta = i_c(q - \psi)$.

Independent parameters in the equations (17.5), taking into account (17.1, 17.4), are q, ϑ. The angle ψ determines location of the straight line OO_u (Fig. 42) with respect to the axis Ox at the moment of profiling the design point P. In this phase of the machine-tool engagement $\psi = q = 90^\circ + \mu - arcsin(r_0 \cos\beta_c/U)$, where $\mu = arcsin(a_w/R_c)$, $\beta_c = \beta_n + \mu$; $R_c = \left[(R + \Delta R)^2 + a_w^2\right]^{0.5}$, β_c, R_c is the helix angle and the mean cone distance of the generating gear, respectively.

Letus consider an influence of the tooth machining scheme on the longitudinal tooth shape, using the tooth section by the plane, normal to the tooth line.

Letus consider as an example, the results of the gearing analysis, the main parameters of which are as follows: number of teeth $z_1/z_2 = 9/31$, external tangential module $m_{te} = 10$mm; mean helix angle $\beta_n = 35^\circ$; direction of the tooth line – left on the pinion, right on the gear; axes angle $\Sigma = 90^\circ$; basic rack profile – according to the standard GOST 15023-76; gear toothing width $b_w = 0.3R$. The nominal diameter of the cutter head is $d_0 = 2286$ mm. Different combinations of the center distance a_w in the machine-tool engagement and the axial setting $\Delta A = \Delta R/cos\delta$ of the gear have been analyzed.

It is supposed that the gear tooth cutting is realized by the two-side method by the cutter head having generating radii of the external cutters greater than the generating radii of the internal cutters. The cutter head of the pinion "grips" [172] the cutters of the gear head. When inserting one head into another, their generating surfaces touch each other by circumferences, corresponding to lines of action. The calculations showed that the gear tooth has a narrowing longitudinal form in case of a conventional cutting method, when $a_w = 0, \Delta R = 0$. The internal normal thickness of the gear tooth in this case is less than the external one approximately by 3mm. This leads to the appearance of "ridges" at the pinion root near the external face.

The curve 1 in Fig. 45 and curves 1, 2 in Fig. 46 show the change of the normal tooth thickness and the helix angle along the gear rim at the conventional scheme of tooth machining. Taking into account the center distance in the machine-tool engagement, when increasing the helix angle of the generating gear, the gear helix angles increase at the external face and decrease at the internal one, in comparison with the previous variant (curves 3, 4 in Fig. 46). This results in the equalization of the pitch normal tooth thickness: at the internal face it increases by *(2...4)%*, and at the external one it decreases by *(10...14)%*, that is shown by the curve 3 in Fig. 45. The external normal tooth thickness becomes even less than the mean normal thickness, that is, the longitudinal tooth shape approaches to equally wide form. Because of increasing the internal normal tooth thickness, the tool module can be increased by *(2...4)%*. More positive results can be obtained by combination of shifts a_w and ΔA of the workpiece (the curve 4 in Fig. 45). Taking into account the center distance in the opposite side, when the helix angle of the generating gear is decreased (the curve 2 in Fig. 45), leads to the increasing of the direct narrowing of teeth. In all considered cases, the normal pitch thickness of the mating pinion, cut by enveloping cutters, does not depend on the shift of axes in the machine-tool engagement and almost does not change along the gear rim.

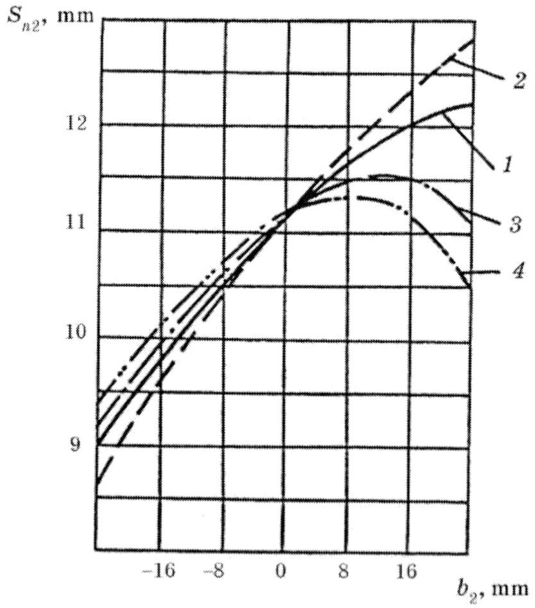

Figure 45. Variation of the pitch normal thickness of the gear tooth along the gear rim: 1 – $a_w = 0, \Delta A = 0$; 2 – $a_w = -30mm, \Delta A = 0$; 3 – $a_w = 20mm, \Delta A = 0$; 4 – $a_w = 20mm, \Delta A = 30mm$.

Thus, by changing the scheme of tooth machining, it is possible to reduce the difference of external and internal normal tooth thickness, decrease tooth thinning at the internal face and increase the tool module or diameter of the cutter head.

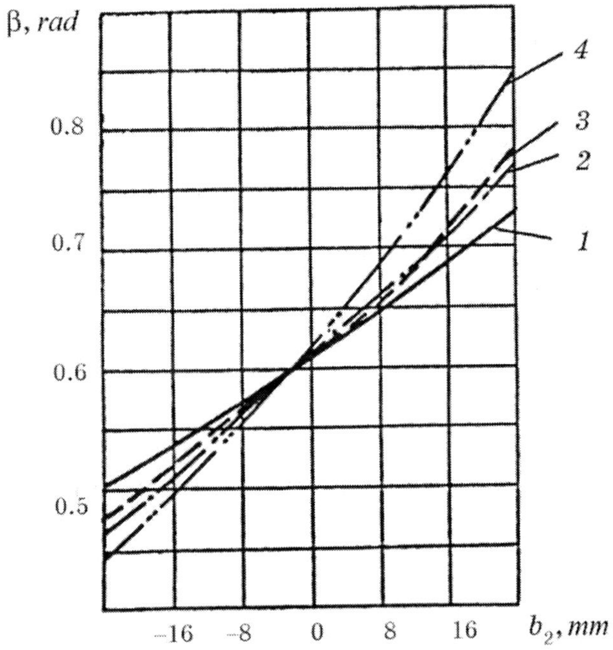

Figure 46. Variation of the pitch helix angle of the gear tooth along the gear rim: 1 – concave side, $a_w = 0, \Delta A = 0$; 2 – convex side, $a_w = 0, \Delta A = 0$; 3 – concave side, $a_w = 20mm, \Delta A = 0$; 4 – convex side, $a_w = 20mm, \Delta A = 0$.

17.3. Main Parameters of Geometry and Quality of Engagement

The load-bearing capacity of bevel NG, as well as of cylindrical one, is determined to a great extent by the basic rack profile of gears. In spite of the considerable design difference of bevel gearing because of mainly insufficient investigation of the problem, the basic rack profiles, developed for cylindrical gears [28, 138], are applied. Due to the appearance of the axial clearance in the operation process; in some cases, a considerable shift (up to $0.1m$) of gears relative to the nominal position is possible. Under these circumstances, good results for the bevel gearing were obtained with a special basic rack profile, composed of the arcs of helices, the curvature of which is changed when moving away from the pitch line to corresponding upper and lower points of the active profile [73], shown in Fig. 47.

Nominal contact points *5* and *14* with a profile angle $\alpha_k = 28°$ are moved away from the pitch line by $0.6m$. The head parts *5-1* and *5-9* are described by arcs of a logarithmic helix $\rho = 0.447681 e^{2.6739\varphi}$ with centers at points O_{5-1} and O_{5-9}. The polar axes $5\text{-}O_{5-1}$, $5\text{-}O_{5-9}$ of coordinate systems, at which the helices are assigned, are drawn through the point *5*. The curvature of helices *5-1* and *5-9* decreases when moving away from the point *5*. The dedendum parts of the tooth basic rack profile *14-9* and *14-18* are composed of a logarithmic helix arcs $\rho = 0.314501e^{4.266\varphi}$ with centers at points O_{14-9} and O_{14-18}. The polar axes $14\text{-}O_{14-9}$, $14\text{-}O_{14-18}$ pass through the point *14*. The addendum height of the basic rack profile is equal to *m*. The convex and the concave parts are conjugated at the point *9* with a profile angle of $12°$, moved away from the pitch line by $0.1m$ towards the addendum. This allows, in comparison with a standard profile [45], for the elimination of the straight-line part with an angle $\alpha \approx 8°$, which is unfavorable according to the tooth contact strength and the tool durability. The ease-off of the tooth active flanks is increased in the area of the gears pitch surface. The maximum profile angle is decreased by $8°$. The conversion curve is described by a circle with a radius *0.29*.

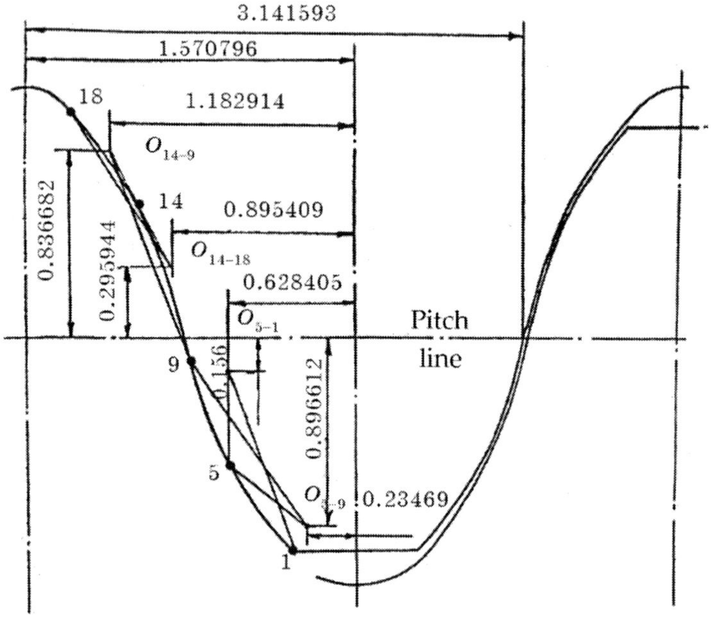

Figure 47. Basic rack profile, generated by arcs of logarithmic spirals ($m = 1$ *mm*).

The influence of a variable curvature of the basic rack profile on the stress for bending was determined by a polarized optical method. The stress in a tooth model of a standard basic rack profile [45] was compared with the stress, determined for the tooth model of the basic rack, having parameters according to Fig. 47. The obtained maximum tension stress on the basic rack profile surface of the model, composed of logarithmic helices arcs, are less than the corresponding stress in the model of the standard basic rack profile by *(14...17)%* due to increasing the curvature radius in the "risky" area. The diagram of stress distribution along the surface of the basic rack profile, according to Fig. 47, has a more pronounced maximum, and point of extremum shifts to the tooth root fillet.

Decreasing the curvature of the tooth normal profile when moving away from the pitch surface at the segment from the nominal contact point to the lower point of the active profile reduces the stress concentration on the concave dedendum. The maximum tension stress area is located, similar to the involute gearing, in the fillet zone, which is not subjected to the compression stress, and sustains only tension stress under one-side loading. The absence of alternate-sign stresses in the "risky" tooth section promotes increasing the metal endurance and fracture strength of gears.

The geometrical analysis of the NG is carried out similarly to the analysis of the bevel quasi-involute gearing with the axial tooth shape III according to standard formulas [38]. The initial data for the analysis along with the parameters of the basic rack profile are as follows: number of teeth z_1, z_2, design module of the gearing (mean normal m_n or external tangential m_{te}), mean helix angle β_n, tooth line direction of the pinion and the gear, axes angle Σ of the gearing, toothing width b_w of the gear, nominal diameter d_0 of the cutter head.

The following peculiarities of NG should be taken into account in this case.: The module and the number of teeth are assigned not only on the basis of the strength balance requirement according to allowable torques from the condition of contact and fracture strength [172] or another required relation between them, but also so that the rational axial overlap ratio ε_β could be obtained along one line of action. The ratio ε_β can be determined from the formula (2.16), which is approximate for the bevel gearing. Here, the design width of the gear toothing is $b_w \le 0.3R$. It is desirable to have such a relationship of parameters z_1, z_2, m_n, β_n, at which the factor ε_β would be equal to $1.1 \ldots 1.2$, and the value of the normal module m_{ni} at the internal face would be equal to the tool module m_0 from the row of standard modules [37]. Diameter of the cutter head d_0 is recommended to be assigned according to the relationship $R\sin\beta_n/d_0=(0.385 \ldots 0.5)$ [138, 172]. In this case, for equal-height teeth the equal-width longitudinal shape [148] is obtained, favorable with respect to the gear strength and manufacturing efficiency of tooth machining, because the maximum tool module is ensured and appearance of "ridges" at teeth roots is eliminated.

However, in practice it is not always possible to apply the heads of such a small diameter because of the secondary cutting action [38].

When choosing the diameter of the cutter head, the expediency of applying the standard head casings and reducing the tooth machining efficiency for decreasing d_0, because of the small number of cutters in the head and low tool stiffness should be taken into account. Small values of heads diameters lead to the mismatch of the casing and the cutters dimensions since, as it was noted, the module is tended to be increased in NG. Standard dimensions of the head casing and the cutter workpiece do not allow to receive the required module for the cutting edge profile, and the produced special cutters have a relatively small base section compared with their height, low stiffness, or it is impossible to place a required number of them into the head of a small diameter. Thus, the technology conditions don't provide the design longitudinal tooth shape only by decreasing the cutter head diameter for a rather large and widely applied combination of tooth numbers [149]. The diameter of the cutter head has to be assigned according to the limitation of the secondary cutting action, and an inefficient single-side cutting method has to be applied for decreasing the longitudinal tooth narrowing. That is why in order to lessen the disadvantages, inherent to the axial shape III of teeth, it is possible to apply specially modified gear cutting scheme [235], which was described above.

Improvement of the engagement quality can be achieved in some cases by the correction of the basic rack profile when cutting the gears, but at the same time it is necessary to prevent the undercut and the thinning of teeth, and also the unallowable shift of the pitch point line in the gearing [129].

Letus consider these factors in details.

Theoretical investigation of the three-dimensional gearing tooth undercut is usually connected with studying the surfaces properties and determination of conditions under which singular points appear, where regularity of the active flank is broken [16, 63, 146]. The position of a plane tangent to the surface is not determined for singular point, as the so-called surface self-intersecting takes place. It is known from the differential geometry, that when writing the surface equation $\bar{r}=\bar{r}(q,\vartheta)$ in a vector form, the condition of singular points existence is described by a vector product $\bar{r}_q \times \bar{r}_\vartheta = 0$. Application of this method allows to determine, whether the tooth undercut occurs, but it doesn't always satisfy the practical requirements. For example, when carrying out the strength analysis, it is desirable to know the depth of the tooth undercut and which part of its active surface by height and length will be excluded from operation as the result of undercut. Besides, the conversion near-pitch-point area of NG teeth (weak by the contact strength) is excluded from operation deliberately in order to avoid the damage in the beginning of operation; that is why a certain undercut of it in a number of cases can be allowed.

Taking into account the noted statements, a numerical research method of undercut is developed [234], which is based on the analysis of a mathematical model of the tooth machining process. At first, coordinates of points of a surface, enveloping a family of tool generating surfaces, are determined from the machine-tool engagement equation $\bar{V}\cdot\bar{n}=0$ according to given equations of the generating surface and the law of relative motion of the tool and the gear workpiece. The values of these coordinates are stored so that the points array would be uniformly located at the necessary part of the tooth side, and the distance between the nearest points would not exceed the required calculation accuracy. Then the surfaces of the tool and the envelope are "rotated" around the corresponding axes, assigning a relative motion, identical to the run-in in the real tooth machining process. Periodically, after rotation of surfaces to a small angle, which is assigned according to the necessary calculation accuracy, the location of all array points of the tooth surface envelope is checked with respect to the generating one, composing their equations in the common coordinate system. The points of the envelope, which appeared to be inside the generating surface, will be located in the area of undercut. Maximum penetration of the envelope point into the tool surface in the process of run-in simulation characterizes the depth of tooth active flank undercut in the vicinity of this point.

Calculations of the depth and the size of the undercut area of bevel NG teeth with a standard basic rack profile [45] are carried out in a wide range of variable parameters for a conventional scheme of tooth cutting.

Letus consider the influence of separate parameters on undercut by an example of one gear. The initial data of calculation results are given below: $z=9$; number of teeth of the generating gear $z_c=45$; external face module $m_{te}=10mm$; $a_w=0$; $\Delta R=0$; $x_\tau^*=0$; $m_0=6\,mm$. For the helix angle $\beta_n=25^\circ$ and diameter of the cutter head $d_0=R/0.7$ the beginning of tooth undercut at the most "risky" point at the addendum on the internal face and with the minimal profile angle, is observed at the shift factor $x_n^*\geq 0.115$ for the gear with a width of toothing $b_w = 0.25R_e$. Increase of the toothing width up to $b_w = 0.3R_e$ leads to a certain reduction of the interval, at which there is no undercut. In this case, the undercut is observed when $x_n^* \geq 0.108$. A similar result is obtained at a negative shift $x_n^* < 0$. With the increase of the toothing width, the depth Δ of undercut is also increased. Thus, for example, at the value of the shift factor $x_n^* = 0.2$ and $b_w = 0.25R_e$ the maximum value of the depth $\Delta = 0.054$ *mm* is changed up to $\Delta = 0.075$ *mm* at $b_w = 0.3R_e$. Further, the results for gears with the toothing width $b = 0.3R_e$, better disposed to undercut, are described.

Undercut can be caused to a small extent by the reduction of the cutter head diameter. At $d_0 = R/1.2$, the undercut begins from the value $x_n^* \geq 0.1$ (instead of $x_n^* \geq 0.108$ for $d_0 = R/0.7$).

Letus compare the calculation results of undercut for gears, differing only by the helix angle. At $\beta_n = 25^\circ, x_n^* = -0.3, d_0 = R/0.7$ the gear having $z=9$ (the rest initial data are described above) is undercut on the internal face for a maximum depth $\Delta = 0.085$ *mm* (the point with a smaller profile angle), and besides, along the height in this section the undercutting is extended to *0.79* mm, that compiles *17.6%* of the addendum active part height. The area, corresponding to the change of the profile angle on the basic rack profile $7.66^\circ \leq \alpha \leq 14.24^\circ$, is undercut on the internal face. Here, the undercut is spread over the whole width of toothing in a range $7.66^\circ \leq \alpha \leq 9.56^\circ$, and for $9.56^\circ \leq \alpha \leq 14.24^\circ$,the undercut area gradually reduces with the rise of the angle α, and its boundary moves away from the external face.

Similar gear with the angle $\beta_n = 45^\circ$ is undercut to a small extent. The obtained results are as follows: $\Delta = 0.049$ *mm*, undercut height *0.64 mm*, or *14.3%*, profile angle of the undercut at the internal face $7.66^\circ \leq \alpha \leq 12.84^\circ$. Thus, when changing the angle β_n from 25° to 45°, the tooth surface undercut area is changed along the height by *3.3%*.

The calculation results are reflected in the diagrams of Fig. 48, where undercut parameters of gears with different number of teeth z and shift factors x_n^* are shown for the helix angle $\beta_n = 25^\circ$, operating width of toothing $b_w = 0.3R_e$, diameter of the cutter head $d_0 = R/0.7$, $a_w = 0$, $\Delta R = 0$, $m_0 = 0.6m_{te}$, $z_c = 45$ at $z \leq 31$ and $z_c = 70$ at $50 \leq z \leq 60$. The shift factors of the basic rack profile are shown horizontally, and the maximum profile angle ϑ_{ni} of the basic rack profile, up to which the corresponding points of the tooth side at the internal face are undercut, is shown vertically. For example, it follows from Fig. 48, that at $x_n^* = -0.4$ for a gear with the number of teeth $z = 20$ all points on the internal face, corresponding to the angle $\alpha \leq 13^\circ$ at the basic rack profile, are undercut, and for the gear having $z = 5$ all points with $\alpha \leq 21.9^\circ$ are undercut.

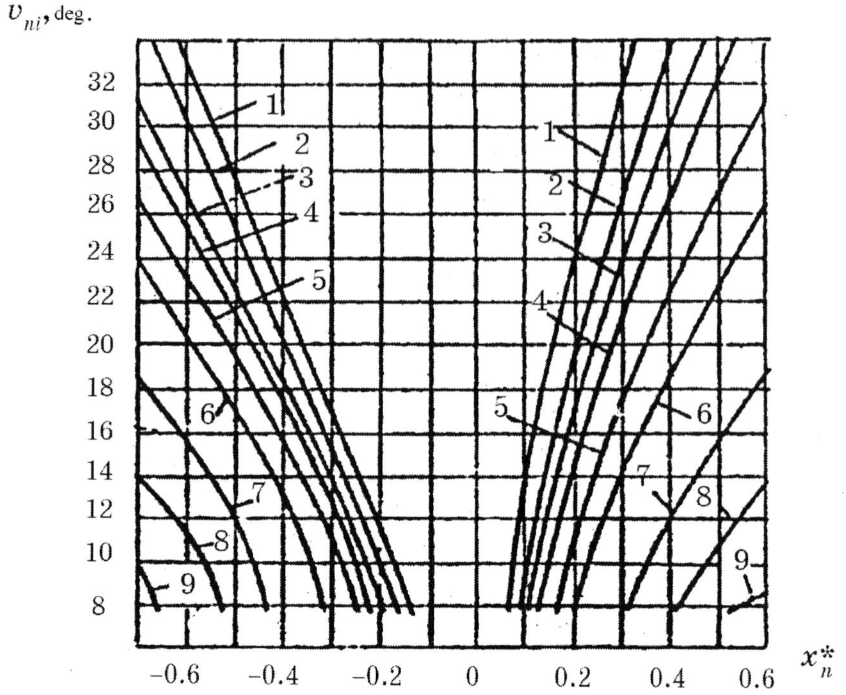

Figure 48. Ultimate angle of undercut at the inner face surface (basic rack profile according to the standard GOST 15023-76): 1 – *z=5*; 2 – *z=7*; 3 – *z=9*; 4 – *z=11*; 5 – *z=15*; 6 – *z=20*; 7 – *z=31*; 8 – *z=50*; 9 – *z=60*.

In the majority of cases, high calculation accuracy is not required to determine the undercut, that is why an approximate method can be applied, replacing the bevel gears by biequivalent spur ones, the number of teeth of which for the internal face is determined as $z_{vni} = z/(\cos\delta \cos^3 \beta_{ni})$ [74]. Then, applying the formulas (3.5), the limits of the basic rack profile shift factors, beyond which the undercut appears, are [128]:

$$\begin{cases} (x_n^*)_{min} = t_2' t_3' [x_a^* - 0.5\rho_a^* \sin\alpha_p (L+1)]; \\ (x_n^*)_{max} = t_1' [x_a^* + 0.5\rho_a^* \sin\alpha_p (L-1)], \end{cases} \tag{17.6}$$

where t_1', t_2', t_3' are correction coefficients, accepted for the gearing with the most widely spread basic rack profiles from Table 28; L - see (3.5).

The error of calculation results of the factors x_n^* according to the approximate formulas (17.6), estimated by means of the technique described above, within the range $9 \le z_{vni} \le 80$, is less than *3%*.

Biequivalent gears can also be applied for an approximate calculation of the tooth thinning, as the exact solution of the problem on active addendum surface intersection with the bevel tips surface in order to determine the tooth thickness S_a^* on the tips surface is rather bulky and not always reasonable.

Table 28. Values of correction coefficients $t_1^{'}, t_2^{'}, t_3^{'}$.

Correction coefficient	Value of the coefficient for a gearing with the basic rack profile		
	RGU-5	According to the standard GOST 30224-96	According to the standard GOST 15023-76
$t_1^{'}$	1.35	1.56	1.6
$t_2^{'}$	1.12	1.3	1
$t_3^{'}$	1	1	$0.42z_{\vartheta ni}^{0.283}$ (for $z_{\vartheta ni} > 40$ $z_{\vartheta ni} = 40$ must be accepted)

The carried out calculations on determining the thinning boundaries of the bevel gearing with the noted above basic rack profiles allowed to establish that thinning is not a limiting factor at all for the gearing with the basic rack profile according to standard GOST 15023-76, and for other types of gearing the limiting value of the factor $(x_n^*)_{max}$ can be determined with sufficient accuracy according to (3.11) and Table 7 (see Chapter 3).

Fig. 49-51 show lines of tooth undercut and thinning boundaries. Exceeding the shift factor by module leads to undercut. It can also be seen from Fig. 49-51, that the tooth thinning restricts the shift factor only for the gearing with the basic rack profile RGU-5 with a low tooth number. For the gearing with the basic rack profile according to the standard GOST 30224-96, within the whole range of tooth number z_{vni} .undercut precedes thinning.

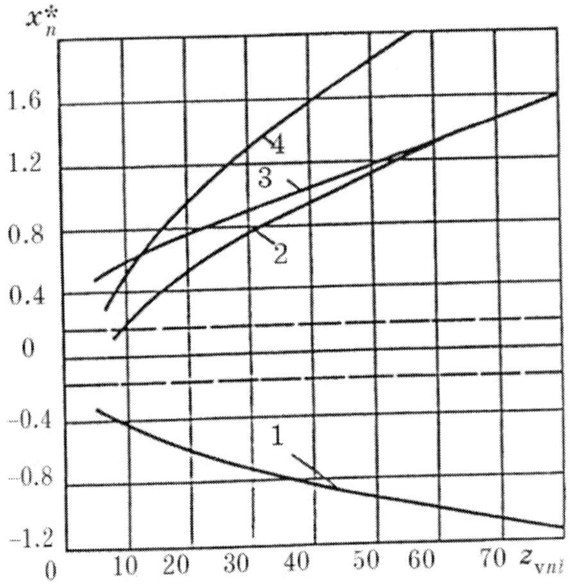

Figure 49. Blocking contour for gears with the basic rack profile RGU-5: 1, 3 – lines of undercut; 2, 4 – lines of thinning of teeth with surface hardening and homogenous structure of material correspondingly.

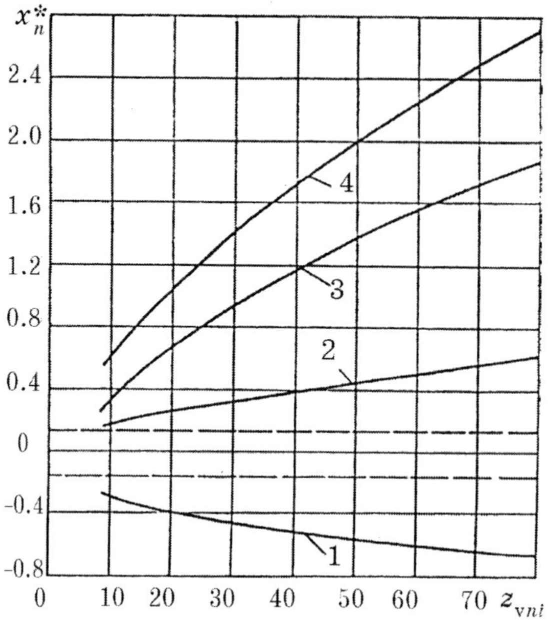

Figure 50. Blocking contour for gears with the basic rack profile according to the standard GOST 30224-96: 1, 2 – lines of undercut; 3, 4 – lines of thinning of teeth with surface hardening and homogenous structure of material correspondingly.

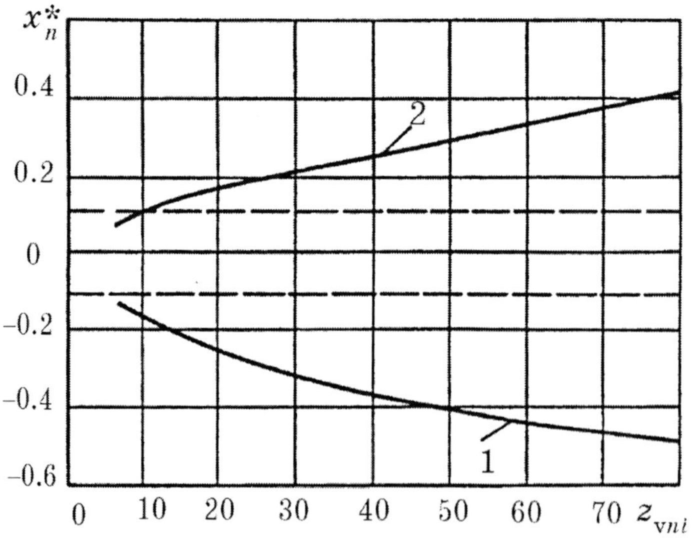

Figure 51. Blocking contour for gears with the basic rack profile according to the standard GOST 15023-76: 1, 2 – lines of undercut.

Dotted lines (see Fig. 49-51) show the limits of the allowable position of the pitch point line in an equally shifted gearing. These limitations are, connected with expedience of excluding the near-pitch-point area, unfavorable according to the contact strength from operation. Remember (see Chapter 3) that for the gearing having the shift in engagement the situation is possible, when the pitch point line will intersect the active surface of the tooth addendum or dedendum. Such condition, as it was noted, contradicts the main idea of increasing the contact strength by means of application of the out-pitch-point NG. That is why it is also desirable to limit the basic rack profile shift factor according to the location of the pitch point line. In this case, application of the biequivalent gears for the bevel gearing allows to use the same formulas, as for the cylindrical one. Since the bevel NG is designed, as a rule, equally shifted (the sum of the shift factors for the pinion and the gear is equal to zero), then the condition, preventing the unallowable shift of the pitch point line, is transformed to the form

$$\left|x_n^*\right| \le \left\{t_a^*, t_f^*\right\} \tag{17.7}$$

The plotted blocking contours to the determination of the limiting values of the basic rack profile shift factors from the conditions of undercut and thinning absence, and also of the allowable location of the pitch point line, permitted to design the bevel gearing more reasonably.

An important task, promoting the engagement quality improvement, is the design of a real gearing, produced with certain manufacturing errors, on the basis of engagement kinematics analysis, considered below.

17.4. KINEMATICS OF GEARING WITH TWO LINES OF ACTION AND PREDICTION OF THE CONTACT PATTERN

TLA gearing, and some others [262, are characterized by the presence of two or several lines of action, influencing the interaction of tooth active flanks. This peculiarity of the engagement required a considerable modification [227, 230] of analysis methods of the engagement kinematics, applied for OLA gearing.

To determine a common contact point of two active flanks, a system of equations is solved, representing the coordinates equality (17.5) of two links surface points and unit normal vectors to surfaces at these points:

$$\begin{cases} \bar{r}_1 = \bar{r}_2; \\ \bar{n}_1 = \bar{n}_2. \end{cases} \tag{17.8}$$

The system (17.8) involves five independent equations and six unknowns. Along with the noted parameters q, ϑ of each surface, the parameters, determining rotation of kinematic links, to which these surfaces belong, that is, in our case angles φ_1, φ_2 of the pinion and the gear rotation in operating engagement, should be referred to the unknowns. Consequently, assigning the values of one of angles φ_1 or φ_2, values of the rest parameters are found, coordinates of common contact points, line of action, contact lines and angles of rotation of the second link in each phase of engagement are determined. If other interacting pairs of active flanks (areas) of these teeth are considered in the same way, then with each pair of surfaces, the system will be increased by five equations, containing four unknowns. Angles

φ_1, φ_2 must be equal for all pairs of surfaces because of the rigid connection of teeth parts between each other. The number of equations in the system will exceed the number of the unknowns. It will appear, due to the unreasonable introduction of redundant constraints between parameters, assuming simultaneous interaction of a pair of teeth in several common contact points, which can occur only at specially chosen active flanks, their ideal manufacture and at nominal location of rotation axes. In a real gearing, it is impossible to determine before the calculations which pairs of surfaces exactly contact in the necessary phase of engagement. In order to eliminate this contradiction, it is proposed to carry out kinematics analysis of the gearing with several lines of action in the following way [230].

Letus compose k systems of equations (17.8), corresponding to k pairs of contacting surfaces at k lines of action. Each system is solved separately, conditionally assuming independence of each pair of surfaces interaction. Here, assigning a certain value of the angle of one of the gearing link rotation, for example, of the driving one - φ_1, k values of the driven link angle $(\varphi_2^1, \varphi_2^2, \varphi_2^3, ...\varphi_2^k)$ of rotation, corresponding to it, are determined. The actual angle φ_2 of rotation of the driven link is determined according to the condition of mutual penetration absence of interacting surfaces. For the driven link, this angle will correspond to maximum of the calculated values $\varphi_2 = max\{ \varphi_2^1, \varphi_2^2, ...\varphi_2^k \}$. If several equal to maximum angles from k angle values are resulted from calculation, it will mean that several pairs of surfaces interact simultaneously in the phase of engagement corresponding to the angle φ_1.

The error $\Delta\varphi_2$ of the angle of the driven gear rotation is determine by , comparing the calculated value of the angle φ_2 with its nominal one. The difference of angles φ_2 and one of the angles $(\varphi_2^1, \varphi_2^2, ...\varphi_2^k)$ determines the angle, to which the driven gear should be rotated to ensure the interaction between the corresponding pair of contacting surfaces in the given phase of engagement. Thus, it is possible to determine the minimum clearance between any pairs of contacting surfaces in the given phase of engagement. The last circumstance is essential for the strength analysis. Determining clearances and location of common contact points on the areas of teeth surfaces, not only the absence of kinematic interaction of some areas is stated, but also the basis for prediction of the load redistribution between separate pairs of areas as a result of tooth deformation is obtained.

Letus consider some practical problems, appearing when realizing the described approach.

Two scalar equations correspond to a vector equality $\bar{n}_1 = \bar{n}_2$. Formally, it doesn't matter which two of three projections of the unit normal vector will be equated. But it is more reasonable to keep the projections of the normal to axes of the pinion and the gear rotation, which do not change at their rotation. The first equation, depending only on the angle φ_1, and the second – only on the angle φ_2, are obtained, that makes possible to reduce the system (17.8) to a solution of three scalar equations, corresponding to the vector equality $\bar{r}_1 = \bar{r}_2$.

Indeed, taking into account (17.5), we can find $\bar{n}_i = \frac{\partial \bar{r}_i}{\partial q} \times \frac{\partial \bar{r}_i}{\partial \vartheta}$ and projections of the unit normal vectors in coordinate systems S_i, related to the cut gears: $n_{xi}(q_i, \vartheta_i)$; $n_{yi}(q_i, \vartheta_i)$; $n_{zi}(q_i, \vartheta_i)$.

Taking into account rotation of the pinion to the angle φ_1, and the gear – to the angle φ_2^k, we have from Fig. 43: $n''_{y1}(q_1, \vartheta_1, \sin\varphi_1, \cos\varphi_1) + n''_{z2}(q_2, \vartheta_2) = 0$, $n''_{z1}(q_1, \vartheta_1) - n''_{y2}(q_2, \vartheta_2, \sin\varphi_2^k, \cos\varphi_2^k) = 0$.

Making transformations, similar to those carried out when solving the equation (17.3), we will obtain $\varphi_1 = \varphi_1(q_1, \vartheta_1, q_2, \vartheta_2)$; $\varphi_2 = \varphi_2(q_1, \vartheta_1, q_2, \vartheta_2)$.

Substituting values of angles φ_1, φ_2 into (17.8), we will reduce (17.8) to a system of three transcendental equations:

$$\begin{cases} f_1 = x_1(q_1, \vartheta_1, q_2, \vartheta_2) - x_2(q_1, \vartheta_1, q_2, \vartheta_2) = 0; \\ f_2 = y_1(q_1, \vartheta_1, q_2, \vartheta_2) - y_2(q_1, \vartheta_1, q_2, \vartheta_2) = 0; \\ f_3 = z_1(q_1, \vartheta_1, q_2, \vartheta_2) - z_2(q_1, \vartheta_1, q_2, \vartheta_2) = 0. \end{cases} \tag{17.9}$$

The system (17.9) has the infinite number of solutions, corresponding to each point of the line of action. To find one of these solutions, one of the parameters, for example, q_2, should be assigned by arbitrary numerical values, and the obtained normal system of transcendental equations should be solved by a known method of variation of parameters [129]. As the result of solution of corresponding systems of equations (17.8), parameters of contacting points and values of gear rotation angles in operating engagement of the gearing with k lines of action will be obtained.

To implement this method, the algorithm and the computer program have been developed for TLA gearing. The initial data includes information about the main geometrical parameters of the gearing, machine-tool settings for cutting the pinion and the gear, and characteristics of the tooth cutting tool. During calculation, according to the given tool parameters and machine-tool settings, the equations of gears tooth flanks are determined in a parametric form. Then the process of the teeth touching is simulated for an operating engagement without load, taking into account possible technological errors at the fixed position of the driving pinion in a definite phase of rotation, set in a cycle with a required pitch. Coordinates of the points of touching in the noted phase of rotation, determined by angle φ_1 of the pinion rotation, are found separately for the points approaching the pitch point area and the points of after the pitch point area. The angle $\Delta\varphi_2$ of the additional gear rotation is determined from the position, corresponding to a uniform rotation, to a flank touching, and according to the calculation results, diagrams of relationships between the angles $\Delta\varphi_2$ and the angle φ_1 of the pinion rotation, when the pinion addendum interacts with the gear dedendum and the pinion dedendum – with the gear addendum, and diagrams of function of the contact point location along the width and the height of the pinion toothing at variation of the angle φ_1, are drawn. To clear up the influence of the neighboring pairs of teeth, the obtained lines are shifted to the value of the angular pitch.

A statement of the theory of engagement, according to which the NG, formed by the principle of a rigid non-congruent generating pair can (regardless the tooth deformation under load) transmit the motion "ideally", with a zero kinematic error, along two lines of action simultaneously only at a nominal position, is proved by calculations. When changing the mutual position of the tooth flanks because of gearbox elements deformation or at existing errors of assembly and manufacture of gears, motion of the driven gear becomes non-uniform, active contact lines are reduced, teeth transmit the load along one contact area instead of two theoretical ones (except separate moments of the tooth changeover). Because of concentration and the dynamical component of the load, the load-bearing capacity is considerably decreased [132].

Fig. 52 shows a diagram of the contacting of the gearing with parameters $z_1 = 9$, $z_2 = 31$, $\beta_n = 35^\circ$, $m_{te} = 10\,mm$, $r_0 = 114.3\,mm$, basic rack profile according to the standard GOST 15023-76, cut by a theoretically precise scheme with crossed axes of machine-tool engagement. It is supposed that the tool and the machine-tool are "ideally" precise, the gearing is mounted in the gearbox without errors. The exception is the non-intersection of axes in the gearbox, equal to *0.1 mm*. In this case, as it is seen from Fig. 52, in a gearing, which is "ideally" kinematically conjugate along two lines of action at a nominal position, at uniform rotation of the driving pinion the motion of the driven gear becomes non-uniform. The active contact lines are reduced by more than *50%* and are shifted towards the external face. The contacting begins at the pinion dedendum near the external face in a phase of engagement, corresponding to the angle of the pinion rotation $\varphi_1 \approx 0.37\ rad$. Then, until the pinion rotates to the angle $\varphi_1 \approx 0.1\ rad$, the contact point is displaced along the pinion dedendum (line f_1) to the middle of the gear rim, the driven gear slows down its rotation. In this phase of engagement, the tooth changeover from dedendum to addendum of the pinion takes place. Contacting of dedendum at the point located at a distance of *6 mm* away from the middle of the gear rim ($b_1 = 0$) is stopped, and the second contact line (a_1) at the pinion addendum comes into operation. At the moment of the tooth changeover, an instant change of the gear angular position by $\approx 10^{-4}\ rad$ occurs because of a mismatch of position of the pinion and gear active flanks. Due to impact, the angular position of the gear approaches a nominal value, but further, the gear again rotates slowly until the contact changeover from the pinion tooth addendum (at the point, located *3mm* away from the rim middle towards the internal face) to a dedendum of the neighboring pinion tooth. Here, the impact occurs again with an advanced rotation of the gear approximately by $7 \cdot 10^{-5}\ rad$.

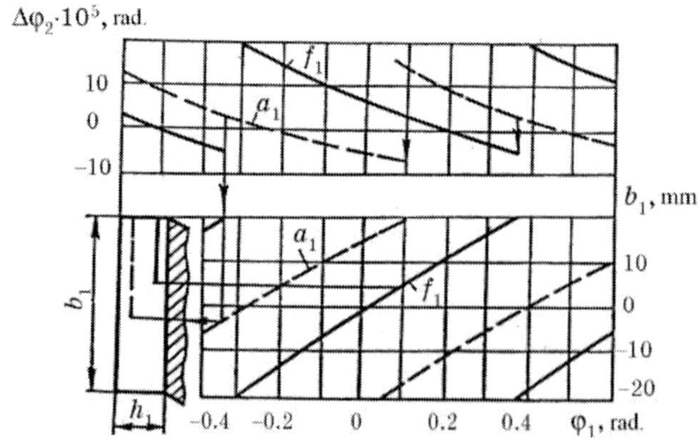

Figure 52. Influence of center distance errors (non-intersection of axes *0.1 mm*) on location of contact points, active contact lines and errors of the gear rotation angle.

Diagram Fig. 52 illustrates the influence of a center distance assembly error on the location along the gear toothing width and along the tooth height h_l of active contact lines and on errors of rotation angles of the driven gear.

Revealing the influence of tooth cutting errors on the noted characteristics is of great interest, that will help a purposeful correction for improving the contact pattern and vibroacoustic factors. As an example in Fig. 53 the influence of the rolling-in gear ratio error on the active contact lines and the error of the gear rotation angle is represented. In this case contacting appears only along one active contact line at the pinion tooth dedendum (gear tooth addendum). More detailed results of investigations are described in works [227, 232].

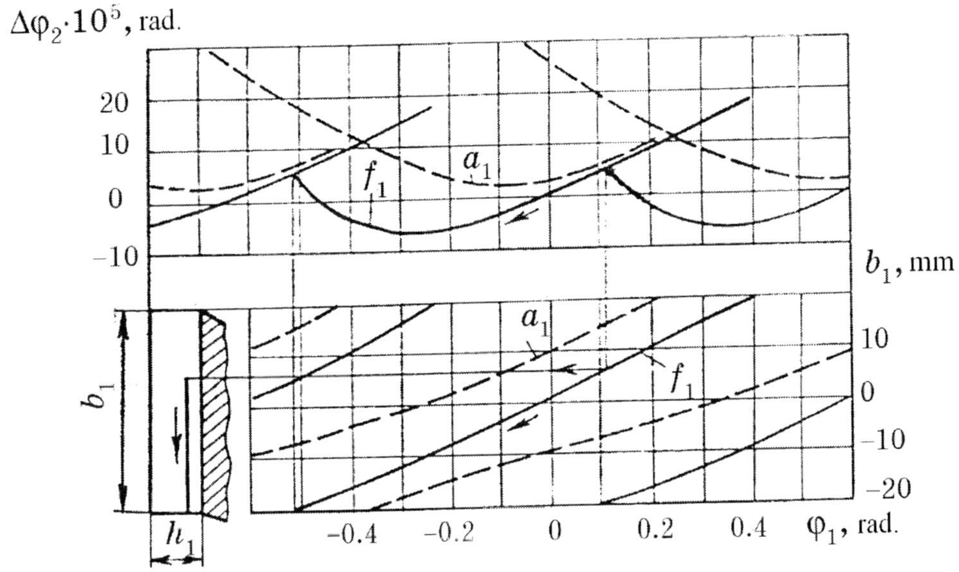

Figure 53. Mismatch of contacting along two lines of action due to tooth-cutting errors ($\Delta i_{cl} = -0.003, a_w = 20mm, \beta_n = 45^\circ$).

The described method allows to determine the location of contact points and active contact lines of TLA gearing, taking into account manufacturing errors, that enables turning to solution of an important practical task – prediction of the contact pattern.

It is known that the shape, location and “behavior” of errors at the contact pattern determine to a great extent the quality of engagement of the bevel pair.

Letus place instant contact areas at points of active contact lines, which will be determined by clearances distribution between the teeth flanks in the vicinity of these contact points. A set of initial contact areas form an integral contact pattern. Letus assume outlines of equal clearance in the vicinity of the point of teeth kinematical touch as an analog, determining outlines of contact areas. The clearance function analysis had been carried out earlier in [220] and a simplified formula was used for calculation. It is proved by a graph-analytic method of determining the equal clearance function and by experiments [17], that for the plastic bevel NG, the contact area shape corresponds to the function of equal clearance between teeth. The noted research was carried out for a gearing, cut by generation method according to the conventional scheme.

It should be noted that in order to determine the direction and size of a touching area of quasi-involute bevel gears, the method of clearance between contacting surfaces is also applied in TCA program [267].Therefore, it is experimentally established that, when checking the gearing "by paint" at the inspection machine, the contact area is formed due to painting of teeth surfaces in the area, where they are apart from each other for less than *0.006 mm.*

To obtain the equal clearance outline in the vicinity of tooth contact points of the pinion and the gear, letus draw a straight line, parallel to direction vector $\bar{n}(n_{xi}, n_{yi}, n_{zi})$ of the common perpendicular to contacting surfaces, through the current point of the tooth flank $P_1(x_1, y_1, z_1)$. The point $P_2(x_2, y_2, z_2)$ of intersection of this straight line with the tooth side of a mating gear will be the required one, if the distance Δ between the points P_1 and P_2 is equal to the given one. A system of three equations with four unknowns is obtained:

$$\begin{cases} f_1 = \dfrac{x_2(q_2,\vartheta_2)-x_1(q_1,\vartheta_1)}{n_{xi}(q_1,\vartheta_1)} - \dfrac{y_2(q_2,\vartheta_2)-y_1(q_1,\vartheta_1)}{n_{yi}(q_1,\vartheta_1)} = 0, \\ f_2 = \dfrac{x_2(q_2,\vartheta_2)-x_1(q_1,\vartheta_1)}{n_{xi}(q_1,\vartheta_1)} - \dfrac{z_2(q_2,\vartheta_2)-z_1(q_1,\vartheta_1)}{n_{zi}(q_1,\vartheta_1)} = 0, \\ f_3 = \sqrt{\begin{aligned}&[x_2(q_2,\vartheta_2)-x_1(q_1,\vartheta_1)]^2 + [y_2(q_2,\vartheta_2)-y_1(q_1,\vartheta_1)]^2 + \\ &+ [z_2(q_2,\vartheta_2)-z_1(q_1,\vartheta_1)]^2\end{aligned}} - \Delta = 0. \end{cases} \tag{17.10}$$

The system of equations (17.10) by assigning fixed values for one of the parameters, for example, q_1, is transformed to a form $f_i(x_1, x_2, \ldots x_n) = 0$ $(i = 1, 2, \ldots, n)$ and is solved by a numerical Newton's method.

A complete pattern of the clearance distribution between the tooth flanks in the vicinity of the contact point is obtained when assigning the number of fixed values of the clearance Δ in the equation f_3.

Boundaries of the contact area are determined (similar to [267]) by a clearance between the tooth flanks in the direction of the common normal, equal to $6\ \mu m$.

When increasing the loads, the distances Δ in the system (17.10) should be respectively increased by the values of the contacting surfaces approaching, obtaining the instant contact areas.

Fig. 54 shows an example of determination of contact and touching areas boundaries at different approaching of surfaces for contacting of the pinion addendum (convex side) with the gear dedendum (concave side), having parameters: basic rack profile according to the standard GOST 15023-76, $z_1 = 10, z_2 = 31, m_0 = 6mm$, $m_{te} = 10mm$, $d_0 = 3048mm$, $\beta_n = 35^\circ$, $a_w = 0$, $\Delta R = 0$. Distances from the pitch surface (zero mark) towards tooth dedendum (with "minus" sign) are marked horizontally, and distances from the gear rim middle (zero mark) to the internal face of the gear (also with "minus" sign) are marked vertically.

It should be notedthat a clearance field in the vicinity of the tooth flanks touching point can be applied as the initial data for a contact problem solution [2].

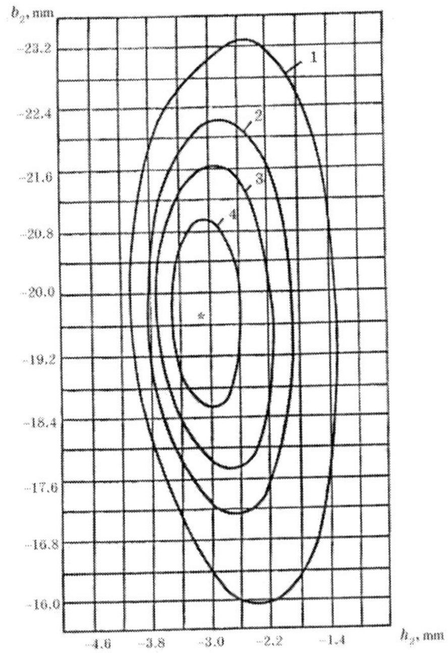

Figure 54. Areas of tangency and contact: 1 – $\Delta = 46.6$ mcm; 2 – $\Delta = 25.3$ mcm; 3 – $\Delta = 16$ mcm; 4 – $\Delta = 6$ mcm (area of tangency).

17.5. Synthesis of Gearing with Two Lines of Action

The task of synthesis of conjugated bevel gearing is determination of requirements, which must be set on generating surfaces and their relative motion, when cutting tooth active flanks, so as the obtained surfaces would become conjugate surfaces of gears [146]. It was mentioned above that modification of tooth flanks of a quasi-involute engagement, leading to the generation of the point contact instead of the linear one, is implemented in practice. Methods of approximate OLA gearing synthesis were thoroughly developed for bevel and hypoid gearing. The synthesis of approximate gearing is inseparably related to a technique of tooth profiling. The synthesis of three-dimensional gearing with bevel gear is understood [148] as the task the machine-tool settings detection, carrying out the tooth cutting of interacting gears, according to the condition of obtaining the required quality of engagement. Methods of local synthesis of a quasi-involute gearing according to different quality criteria were developed [62, 146, 245]. Implementation of the given gear ratio and definite conditions of engagement are ensured at the design point. But even for a quasi-involute gearing, it is practically impossible to satisfy simultaneously all manufacturing requirements, such as the given shape, location and size of contact areas, taking into account contact deformations, "behavior" of the contact pattern and noise reduction. For NG with two or more lines of action, characterized by sensitivity to radial errors, displacement of the contact area along the gear rim and various conditions of the tooth addendum and dedendum contacting, limitation of local synthesis methods is aggravated.

An important forward step was the development of machine-tool settings calculation for cutting a hypoid NG according to several points of one line of action [75]. Certain conditions are set to the contact at the three considered points, one of which can be the equality of instantaneous and mean gear ratios. By solving a system of *24* transcendent equations, the reduction of a cyclic gearing error and the correction of the line of action are achieved. But for two or more lines of action, the exact solution can't be obtained since the number of equations will exceed the number of varying parameters.

In research works [52, 53], a method of non-local synthesis of an approximate quasi-involute bevel gearing according to the points of geometrical changeover is described. In this case, the attempt was made to take into account tooth deformation under load in the first approximation.

Further development led to consideration of the gearing synthesis as a problem of searching optimal values of tooth machining settings [190, 241]. The optimization task of setting parameters of a quasi-involute bevel gearing has been successfully solved [148, 180].

For bevel NG TLA, it is required to create a mathematical model, to develop quality characteristics and criterion of optimality, taking into account peculiarities of this engagement, particularly, mutual influence of lines of action on working capacity at inevitable appearance of manufacturing and assembly errors. Quality of the gearing depends on a number of technical and economical characteristics. In practice, it is required to take into account many contradictory factors, influencing the load-bearing capacity, durability, cost, vibroacoustic characteristics of the gearing and tooth machining efficiency.

One should proceed from the fact that achieving the "ideal" constant gear ratio of the pair without load (or even under a constant load), is not an optimality criterion of the gearing design. The synthesis of gearing having more than one line of action, must be based on synchronization of contacting along all lines of action in the presence of manufacturing errors. The required contact localization will cause the appearance of a certain non-uniform rotation of gears. In this connection, it is important to have the possibility to control the characteristics of engagement for correct gearing design with account of operating conditions, manufacturing and design peculiarities.

For NG with two or more lines of action, from the viewpoint of its strength, it is desirable that points of the pinion and gear, located at all active contact lines, could interact simultaneously in each phase of rotation. In the first approximation, the quality of active surfaces and tooth cutting machine-tool settings can be characterized by synchronization of surfaces touching without load at points of each active contact line and angular position of the gear in the phase of beginning, middle and end of contacting of one teeth pair. It is necessary to provide a nominal gear ratio in the middle of the gear rim, and the driven gears must be rotated to a slightly less angle than at uniform rotation in phases of the teeth changeover. For impact-free operation, the end of engagement of one pinion tooth and the beginning of another one contacting must take place at the same values of the driven gear rotation angle. The desired lag of the gear rotation angle $\Delta\varphi_{kp}$ from the position of uniform motion, determined by the degree of longitudinal contact localization, at points of teeth changeover can be obtained in each specific case according to the required value of deviation δl_T of the actual tooth line from the theoretical one in a face plane and to the gear radius r_2 from the equality $\Delta\varphi_{kp} = \delta l_T / r_2$. Recommended values of the deviation δl_T are described in Chapter 18.

In a general case, there are no combinations of machine-tool setting parameters at which the cut surface satisfies all quality criteria simultaneously [148]. The optimization task within the synthesis of engagement is reduced to obtaining the best approximation to the given quality criteria from a set of possible values of the settings. The basic machine-tool settings for cutting the gear with account of the required longitudinal gear tooth narrowing are defined below (paragraph 17.6). Machine-tool setting parameters for the pinion are found from the condition of obtaining the minimum value of a kinematics optimality criterion of the gear pair.

Corrections of a cutter head radius Δr_0, gear ratio Δi_{c1} of a generation train, radial setting ΔU_1 of the cutter head, setting correction Δa_{w1} of axes in the pinion machine-tool engagement and axial displacement ΔA_1 of a workpiece are considered to be independent parameters for one-side pinion machining. In a limited range of values, the correction of angles of machine-tool stock mounting is also possible. The value of variable limits is determined by the possibility to implement the settings at a definite tooth cutting machine-tool and by the appearance of singular points (for example, during undercut) on teeth surfaces.

With regard to a gearing with two lines of action, letus consider six pairs of contact points, corresponding to three phases of a pair of teeth contacting along each line of action. Let points *1, 2* be located near the internal face, points *3, 4* – near the middle of a gear rim, and points *5, 6* – near the external face, the odd points corresponding to the active line of the pinion dedendum. The value of a kinematic error of the driven gear for contacting at a corresponding point will be found by solving separately two systems of equations (17.8), composed with account of the generating surface geometry for odd and even points. Besides zero error $\Delta\varphi_3 - \Delta\varphi_4 = 0$ of the gears rotation at points *3* and *4*, it is necessary to have the required difference of the gear angles error for contacting at other points. Letus represent a complex criterion of engagement kinematics optimality in the form of a linear combination:

$$F = \sum_{i=1}^{7} c_i f_i(\Delta r_0, \Delta i_{c1}, \Delta U_1, \Delta a_{w1}, \Delta A_1); \quad F = 5(\Delta\varphi_3 - \Delta\varphi_4)^2 +$$
$$+10(\Delta\varphi_1 - \Delta\varphi_5)^2 + 10(\Delta\varphi_2 - \Delta\varphi_6)^2 + (\Delta\varphi_6 - \Delta\varphi_4 - \Delta\varphi_{kp})^2 + \tag{17.11}$$
$$+(\Delta\varphi_2 - \Delta\varphi_4 - \Delta\varphi_{kp})^2 + (\Delta\varphi_5 - \Delta\varphi_3 - \Delta\varphi_{kp})^2 + (\Delta\varphi_1 - \Delta\varphi_3 - \Delta\varphi_{kp})^2.$$

Because of the mathematical model complexity of bevel Novikov gears tooth cutting and engagement, the complex criterion of optimality is a non-linear function of setting parameters, that causes the requirement to solve a task of the function minimization by iterative methods. We used a method of "the quickest descent".

Fig. 55 shows diagrams of the same gearing contacting as in Fig. 52, but after the optimization of the setting parameters. The gearing is cut according to a modified scheme of tooth machining (Fig. 42) with axes shift in machine-tool engagement of the pinion and gear $a_w = 10mm$. Pinion corrections are: $\Delta i_{c1} = 0.00101$, $\Delta a_{w1} = 0.15mm$, $\Delta U_1 = 0.1mm$, $\Delta r_0 = 0.1m_0$, $\Delta A_1 = 0.52\cdot 10^{-4}mm$ (such a small value was calculated by computer program, practically $\Delta A_1 = 0$). At the same error value of the gears mounting into a gearbox, the diagram of kinematic error became "arc-shaped" (Fig. 55) instead of "saw-shaped" (Fig. 52).

The shift of the contact area to the face is excluded, impacts during the teeth changeover are absent. Acceleration and retarding of the gear motion are realized smoothly, which must affect positively the gearing dynamics. The same advantageous type of teeth contacting also takes place when changing the axial setting of gears with respect to the nominal position within limits $\pm 0.1\,mm$, and when changing the sign of the gearing center distance error to the opposite one, in comparison with the one represented in Fig. 55.

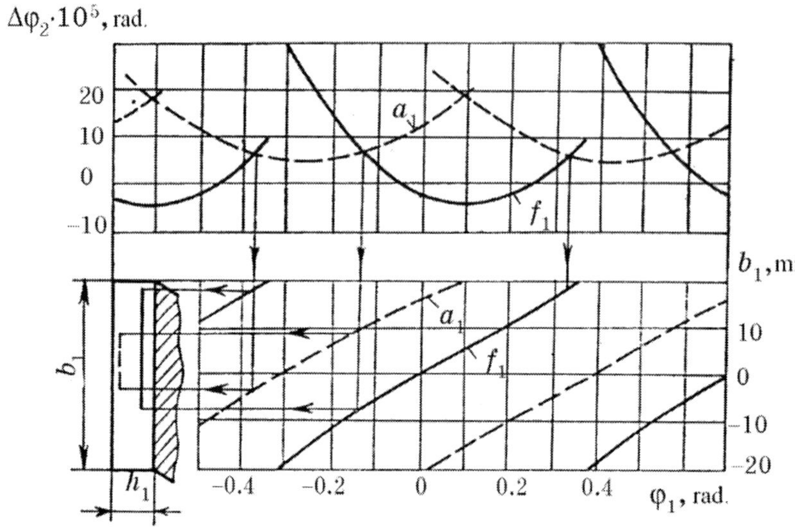

Figure 55. Contacting for non-intersection of axes *0.1mm* in the gearbox of modified teeth of TLA bevel gearing with optimized machine-tool setting parameters for pinion machining.

This example illustrates graphically, the advantages of the modified scheme and reasonability of making corrections of machine-tool settings for cutting the pinion teeth.

17.6. Machine-Tool Settings for Cutting Teeth of the Bevel Pair

When cutting teeth by a simple two-side method, parameters of a generating gear are determined from the relations of Fig. 42. Here, if a relationship $R \sin\beta_n / d_0 = (0.385..0.5)$ is fulfilled, then $a_w = 0$, $R = 0$ are assigned. If the noted relationship is not fulfilled due to increasing of d_0, for example, because of risk of a secondary cutting action, the basic (without account of corrections) center distance a_w in the machine-tool engagement and the shift ΔR are assigned according to the relationship, obtained for $m_n \approx m_{ne}$ [235]:

$$a_w = (0.5\ldots1)\left[d_0^2 - b_w(R+\Delta R) - b_w^2/4 - 2d_0 \sin\beta_n (R+\Delta R)\right]/(2d_0 \cos\beta_n). \qquad (17.12)$$

In cases when it is required to obtain an equal-width [148] longitudinal shape, a_w, ΔR should be assigned so that an equality $d_0 / R_c = (2.0...2.1) \sin \beta_c$ could be fulfilled, where R_c, β_c are the mean cone distance and helix angle of the generating gear, respectively. Then a longitudinal shape will be obtained, for which $m_n > m_{ni}, m_n > m_{ne}, m_{ni} \approx m_{ne}$.

At $a_w \neq 0$ and $\Delta R \neq 0$, the check-up of the gearing for the absence of tooth undercut in the machine-tool engagement is required, which can be carried out by methods described above. (It should be noted that, not only values of shifts $a_w, \Delta R$, but also their relation influence the tooth undercut).

Machine-tool gearing settings for cutting gear teeth are determined from the formulas: radial setting of the cutter head $U_2 = (R_c^2 + r_0^2 - 2R_c r_0 \sin \beta_c)^{0.5}$; angular setting of the cutter head $q_2 = \arcsin(r_0 \cos \beta_c / U_2) - \mu = \pi / 2 - \psi$; axial setting of the head $A_2 = A_2^* + \Delta R / \cos \delta_2$, where A_2^* is the mounting gear distance; shift of the machine-tool table $\Delta B_2 = -h_f - \Delta R \tan \delta_2$, where h_f is the tooth dedendum height; number of teeth of the generating gear $z_{c2} = (z_2 R_c \cos \beta_c)/(R \sin \delta_2 \cos \beta_n)$.

If cutting is carried out by a conventional method, using a rigid non-congruent generating pair, then machine-tool settings for pinion and gear cutting are the same, except for a sign of the head angular setting.

In important power gearing, as it was noted above, the longitudinal localization of the contact pattern is required. That is why when cutting a concave side of a pinion tooth by one-side method, the cutter head radius is increased by the value Δr_{0e}, and for a convex side, it is decreased by the value Δr_{0i}, compared with heads, required to generate a rigid non-congruent generating pair. In the first approximation, it is possible to assign $\Delta r_{0e} = \Delta r_{0i} = 0.1 m_0$. Furthermore, these values are corrected, according to results of the engagement synthesis by a quality criterion, taking into account kinematic characteristics of contacting along two lines of action. In the process of the engagement synthesis, corrections $\Delta U_1, \Delta a_{w1}, \Delta A_1, \Delta i_{c1}$ are also determined for each tooth flank separately, as it is described in paragraph 17.5.

Taking these corrections into account, machine-tool settings for the pinion cutting are

$$U_1 = U_2 + \Delta U_1; \quad a_{w1} = a_w + \Delta a_{w1}; \quad \Delta R_1 = \Delta R + \Delta A_1 \cos \delta_1;$$
$$A_1 = A_1^* + \Delta R_1 / \cos \delta_1; \quad \Delta B_1 = -h_f - \Delta R_1 \tan \delta_1; \quad z_{c1} = z_{c2} + \Delta i_{c1} z_1, \tag{17.13}$$

where A_1^* is the pinion mounting distance.

For gears with the basic rack profile according to the standard GOST 15023-76 for $a_w = 0$, $\Delta R = 0$, a numerical experiment was carried out by computer programs implemented to obtain empirical formulas and, thus, to facilitate the calculation of settings of a pinion with longitudinal tooth modification, ensuring the maximum load-bearing capacity of the gearing. The required value Δd_0 of change of the cutter head diameter for the pinion compared with the head, generating a rigid non-congruent generating pair, and corrections of the pinion settings were determined. Gearing parameters were varied in a range: tool module

$m_0 = (3.15...10)mm$, mean design helix angle $\beta_n = 25^\circ...45^\circ$, gear ratio $u = 1...6.3$, axial overlap ratio $\varepsilon_{\beta n} = 1.15...2.35$, degree of accuracy by smoothness and contact ratings $8...10$ according to the standard GOST 9368-81.

The following relationship has been obtained; characterizing the relation between the change of the cutter head diameter for the pinion cutting and the deviation δl_T of an actual pinion tooth line from a theoretical one [130]:

$$\Delta d_0 = 4\delta l_T \cos\beta_n d_0^2 / \left[(b_w / \cos\beta_n)^2 - 4\delta l_T \cos\beta_n d_0\right], \tag{17.14}$$

Calculations showed that, for $a_w = 0$, $\Delta R = 0$ the corrections $\Delta a_{w1}, \Delta A_1$ and Δi_{c1} improve the quality of engagement to a small extent. In order to obtain a satisfactory contact pattern, smooth teeth changeover and required value of longitudinal modification, along with changing the design diameter of the pinion cutter head, it is sufficient to change radial setting of the pinion cutter head, compared with an "ideally" conjugated gearing.

Radial setting is determined according to the formula

$$U_1 = U_{11} \pm 0.75(U_{11} - U_{10}), \tag{17.15}$$

where $U_{11} = \sqrt{R^2 + 0.25d_{11}^2 - Rd_{11}\sin\beta_{n1}}$; $U_{10} = \sqrt{R^2 + 0.25d_{01}^2 - Rd_{01}\sin\beta_n}$; $d_{01} = d_0 \pm \Delta d_0$ is a design diameter of the pinion cutter head; $d_{11} = d_0 \mp m_0\left[(\rho_f^* - \rho_a^*)\cos\alpha_k + 0.5\pi + l_a^* - l_f^*\right] \pm \Delta d_0$ is a generating diameter of the pinion cutter head at the pitch surface of this head; $\rho_f^*, \rho_a^*, l_a^*, l_f^*, \alpha_k$ are parameters of the basic rack profile according to the standard GOST 15023-76 (see Chapter 1); $\beta_{n1} = \arcsin\dfrac{R^2 + 0.25d_{21}^2 - U_2^2}{Rd_{21}}$ is the pinion helix angle at the pitch surface in the middle of the gear rim; $d_{21} = d_{11} \mp \Delta d_0$ is the generating diameter of the gear cutter head;

radial setting of the gear cutter head

$$U_2 = \sqrt{R^2 + 0.25d_0^2 - Rd_0 \sin\beta_n} \tag{17.16}$$

The upper sign in equations is applied to a concave pinion tooth flank and a convex gear tooth flank, the opposite is for the lower sign.

Letus show as an example, the calculation of the radial setting for cutting teeth of the bevel NG, applied as a main gearing in transmission of a tram KTM-5M, having the following parameters: number of pinion teeth $z = 7$; number of gear teeth 50 ($u = 7.14$); $m_0 = 6$ mm; $b_w = 66\,mm$; $d_0 = 304.8\,mm$; $\beta_n = 23^\circ 16'47''$; $R = 187.926$ mm; $\varepsilon_{\beta n} = 1.214$, basic rack profile – according to the standard GOST 15023-76; degree of accuracy by smoothness rating – 8.

We accept $A_0 = 17.812$ and $b = 0.756$ from Table 36 (Chapter 18), find $\delta_T = 0.069mm$ according to the formula (18.11), get $\Delta d_0 = 4.63mm$ from the expression (17.14) and determine $U_2 = 189.483mm$ according to the formula (17.16); find $U_1 = 190.4\,mm$ for a concave pinion flank, and $U_1 = 188.533mm$ for a convex one according to the formula (17.15). Other setting parameters are determined according to the technique from the work [172].

Fig. 56 shows diagrams reproducing the dependence of the error $\Delta\varphi_2$ of rotation angle of a gear with longitudinal teeth modification ($\Delta d_0 = 4.63mm$) on the angle of pinion rotation, represented by a dimensionless value $\varphi_1' = \varphi_1 z_1 / (2\pi)$ for the cases, when the tooth cutting machine-tool setting corresponds to an "ideally" conjugated gearing (*a*) and when changing of the pinion radial setting (*b*). The diagrams are drawn for the case of contacting of a concave flank of the left pinion tooth with a convex flank of the right gear tooth.

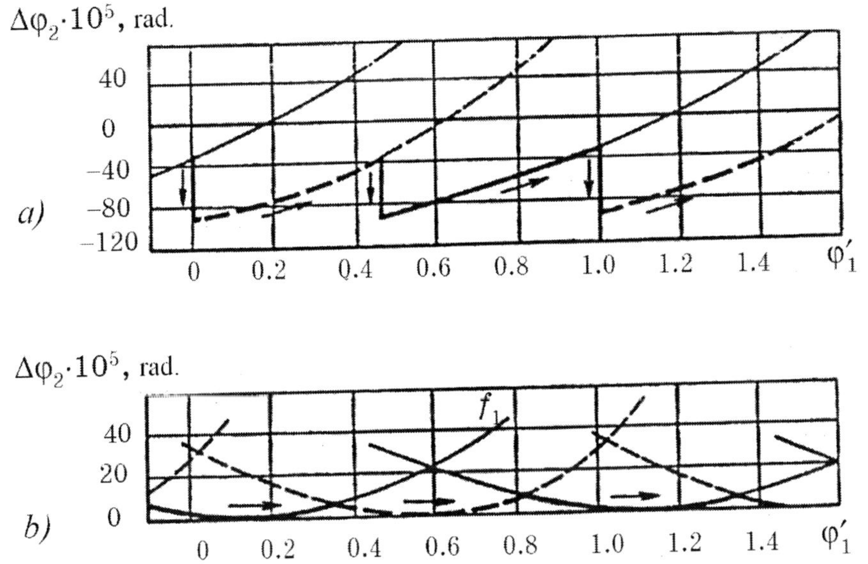

Figure 56. Kinematics of the gearing with longitudinal tooth modification: *a)* – for setting, corresponding to the "ideally" conjugated gearing; *b)* – for corrective adjustment of radial setting of the pinion (solid lines – the contact of the pinion addendum with the gear dedendum; dotted lines – the contact of the pinion dedendum with the gear addendum; arrows – direction of variation of the error $\Delta\varphi_2$ with account of engagement along two lines).

As it can be seen from Fig. 56*a*, at the moment of teeth changeover, the impact occurs at the external face ($\varphi_1' = 0$), which is indicated by an abrupt change of error $\Delta\varphi_2$. The impact at a point located at the external face and gear tooth addendum, "corrects" its angular position. Then the displacement of the contact point from the external face to the internal one, along the gear addendum, takes place with retarding of its rotation. It leads to another impact at $\varphi_1' = 0.46$. Then the gear again rotates slowly, the contact point moves along the gear dedendum from the external face to the internal one until the next tooth begins operating at

$\varphi_1' = 1$, etc. Impacts at rotation cause the overload of the external face during the teeth changeover.

A pattern of the teeth changeover and location of contact points on a tooth is much more favorable at correction with pinion radial setting change (Fig. 56*b*). When $\varphi_1' = 0$, contacting takes place along the dedendum of the previous gear tooth, and only when $\varphi_1' = 0.34$, a smooth changeover occurs. Then the gear at first rotates with acceleration, then with retarding up to a smooth changeover (for $\varphi_1' = 0.82$). The contact points are located approximately in the middle of the gear rim on both lines of action, and the faces are unloaded due to longitudinal tooth modification. In this case, the maximum error of the angle of the gear rotation doesn't exceed $8 \cdot 10^{-5}$ *rad*.

Thus, when determining machine-tool settings for the pinions, formed according to a modified scheme of tooth cutting and with the given longitudinal contact localization, it is necessary to apply special methods of analysis and synthesis of engagement. When the conventional scheme ($a_w = 0$, $\Delta R = 0$) and the basic rack profile according to the standard GOST 15023-76 are used, it is possible to apply approximate simple formulas.

17.7. Peculiarities of Strength Analysis

Two approaches are applied in engineering practice of the bevel gears strength analysis. One of them is based on empirical relations, obtained by processing of long-term test results [254, 265], the other – on reduction of the task to the consideration of equivalent cylindrical gears, constructed on a mean section of bevel ones [74]. A small number of tests of the bevel NG doesn't allow to create a technique of the tooth strength analysis on the basis of experimental results; that is why a technique, applying a scheme of an equivalent cylindrical gear [172], became widespread.

If the mean section of a bevel gear is chosen as a design one, then the principal difference of the bevel gearing from the cylindrical one is ignored - variability of geometrical parameters, determining the tooth strength on the whole along the gear rim, that results in reducing the reliability of the existing techniques. The noted specific characteristic of the bevel gearing predetermines the situation of distribution of the transmitted load and acting stresses,and it is significantly different from the situation for the cylindrical gearing. Therefore, for calculation of stresses in the bevel gearing teeth, it is not sufficient to have information about the load distribution, taking into account the influence of faces (for the cylindrical gearing it is sufficient [100, 110]), but it is necessary to have a complete pattern of the tooth stressed state in different engagement phases in order to determine the most "risky" area. A similar task can be implemented only by means of a computer.

The approach and foundations for making algorithms for a developed computer program CONUS of calculating bending and contact tooth stresses in the bevel NG and some calculation results are described below.

Letus take as a basis the system of equations, representing the elastic equilibrium and consistency of displacements of the tooth contact points for each phase of engagement (10.10), and transform it so that angular parameters could appear instead of linear ones [132]:

$$\begin{cases} \delta W_{\delta j} + \delta W_{kj} + \overline{\Delta}_j + \delta l_j = \delta W_{\delta(j+1)} + \delta W_{k(j+1)} + \overline{\Delta}_{(j+1)} + \delta l_{(j+1)}; \\ \sum_{i=1}^{n} T_i = T_\Sigma; \qquad j = 1,2..n-1. \end{cases} \tag{17.17}$$

It is designated here: δW_δ - total (bending-shift and contact) angular displacement of a tooth at a theoretical contact point; δW_k - angular displacement of a tooth surface, caused by compliance of mating parts (shafts, bearings, etc.); $\overline{\Delta}$ - angular clearance, determined by manufacture and assembly gearing errors; δl - angular clearance, determined by ease-off of tooth contacting surfaces, when they are manufactured with a longitudinal localization.

Let us consider in detail, the components of the system (17.17).

The dependence of a tooth linear displacement on geometrical factors under the action of a normal force for a cylindrical NG TLA (10.11) was obtained in [100, 163]. This dependence can also be applied for a bevel NG TLA with two stipulations: linear displacements must be transformed into angular ones, and for a number of parameters their variable character must be taken into account, that is, the dependence of location ("z" coordinate) of the contact point along the tooth length (such parameters are marked by index "z"). With account of the noted above, the following expression for a total angular tooth displacement can be written:

$$\delta W_{\delta z} = K_{uk} m_0^{1-2x} (0.5 z_{v\Sigma})^a (r_{2z} \cos\alpha_k \cos\beta_z)^{-(1+x)} B_{wz} (\rho^*_{\beta z})^c (K_{bz} T_z)^x, \tag{17.18}$$

where $z_{v\Sigma} = z_1/\cos\delta_1 + z_2/\cos\delta_2$.

It is designated here: K_{uk}, a, x, c - certain constants, depending on parameters of the basic rack profile [100], B_w - parameter, depending on the assigned coefficient x^*_τ of the design tooth thickness change; for the pinion it equals $[0.6 + 0.4(1 + x^*_\tau)^3]^{-1}$, and for the gear it can be calculated according to a relation $\{0.6 + 0.4[2m_n R_z \cos\beta_z /(m_0 R \cos\beta_n) - 1 - x^*_\tau]^3\}^{-1}$; ρ^*_β - longitudinal main reduced radius of surfaces curvature at the contact point, divided by the mean normal module m_n, determined by methods of differential geometry [230].

A subscript "z" indicates, as it was noted, variable parameters changing along the gear toothing width. If the constant part A is marked out, then the formula (17.18) can be written in the form

$$\delta W_{\delta z} = A f(R_z), \tag{17.19}$$

where R_z is the cone distance up to a design point, determining values of the current helix angle β_z, the reduced radius $\rho_{\beta z}$ of surfaces curvature, the pitch radius r_z of the gear, the coefficient K_{bz} of the face influence on distribution of loads and other variable values, comprising (17.18).

Displacement δW_k is determined according to the known formulas [206, 219] with reducing the obtained results to a design point by means of total angular compliance matrix:

$$\delta W_{kz} = [a_{ij}^{*}]_z \{T_i\}_z, \tag{17.20}$$

where $[a_{ij}^{*}]_z$ is the square matrix of total angular compliance for a unit load.

According to the analysis when calculating the angular clearance $\bar{\Delta}_z$, at each theoretical point of engagement, one should take into account: deviations of tangential pitches and profiles of the teeth, axial displacements of the gearing, deviations of the center distance and axes angle of the gearing. In cases when definite values of the noted errors are unknown beforehand, it is reasonable to use calculation of a probable (statistical) clearance on the basis of corresponding tolerances for the given degrees of accuracy according to standards GOST 1758-81 and GOST 9368-81 and the chosen "risk level" (see Chapter 5).

The clearance δl_z depends on the chosen cutter heads radii difference for cutting tooth flanks of the pinion and the gear in order to obtain longitudinal localization. Influence of the latter on the SSS of teeth is considered below (Chapter 18). Here, a case of non-modified teeth ($\delta l_z = 0$) is described.

The solution of the system (17.17) is the determination of coefficients K_{Tz}, and then, similar to the cylindrical gearing (see Chapter 10), of coefficients $K_{H\alpha z}$ and $K_{F\alpha z}$. In this case determination of a theoretical number of contact points at each considered engagement phase, and of coordinates "z" of their location along the gear toothing width can be carried out using the system (17.5).

Acting stresses at different engagement phases of the bevel gearing are determined according to the formulas [132]:

$$\text{contact } \sigma_{Hz} = 2393 K_{Hbz} [K_{Tz} T K_{Hv} /(r_z \cos\alpha_k \cos\beta_z)]^{0.687} (Z_l l_z^{*})^{-1.063} (\rho_{\beta z}^{*})^{-0.312}; \tag{17.21}$$

$$\text{bending } \sigma_{Fz} = 1000 Y_{vz} Y_{\alpha z} K_{Tz} T K_{Fbz} K_{Fv} B_{\sigma z} /(r_z \cos\beta_z m_0^2). \tag{17.22}$$

The parameter $B_{\sigma z}$, taking into account the change of the design tooth thickness along the gear toothing width, is accepted equal to $(1 + x_{\tau}^{*})^{-2}$ for the pinion, and $[2 m_n R_z \cos\beta_z /(m_0 R \cos\beta_n) - 1 - x_{\tau}^{*}]^{-2}$ for the gear.

If the influence of the face on effective contact stresses was taken into account only for extreme (face) location of theoretical contact point (10.18) during engineering calculations of the cylindrical gearing, then for the bevel gearing, calculated by computer programs, it became possible to take into account the influence of the face for any location of the contact pattern, using the relation [132]:

$$K_{Hbz} = (0.5 C_{\alpha\beta z} \beta_z^{0.4})^{1 - t_z / a_{Hz}}, \tag{17.23}$$

where a_{Hz} is the dimension of the major semi-axis of the contact ellipse; t_z is the distance from the design point (center of the contact ellipse) to the gear face; it is obvious, that for $t_z \geq a_{Hz}$ we have $K_{Hbz} = 1$.

Having obtained the stress values from (17.21), (17.22) at all design points of the considered engagement phases, it is possible to find maximum stresses, which determine "risky" points.

Letus give an example of calculations according to the program CONUS of maximum (for all engagement phases) stresses – bending ones for the pinion σ_{F1max}, for the gear σ_{F2max} and contact ones σ_{Hmax}.

Letus consider as an object, the main bevel gearing of a tram KTM-5M (its geometrical parameters are given in paragraph 17.6), gears of which are made of case-hardened steel 12H2N4A, the degree of accuracy is *9*. The nominal power of an electric motor is *45* kW, the rotational frequency of the pinion is $1200\ min^{-1}$.

The results of calculation of maximum bending and contact stresses according to known [172] and proposed techniques are shown in Table 29.

Table 29. Comparison of tooth stresses of the tram KTM-5M main gearing, determined by known [172] and proposed techniques.

Parameter	Stresses (MPa) by the technique		
	according to [172]	proposed (according to the program CONUS)	
		for a real gearing	for a gearing with $\bar{\Delta}_z = 0$ and $\delta W_{kz} = 0$
σ_{F1max}	90	288	135
σ_{F2max}	55	225	180
σ_{Hmax}	510	1058	580

Maximum stresses, determined according to the program CONUS for the described gearing without errors ($\bar{\Delta}_z = 0$), and without influence of compliance of mating parts ($\delta W_{kz} = 0$), are shown for comparison.

As it is seen from Table 29, values determined according to the technique [172], are considerably smaller, than ones obtained according to the proposed technique, even including the case when $\bar{\Delta}_z = 0$ and $\delta W_{kz} = 0$. It is explained, as it was noted above, by the fact that the analysis [172] is based only on the mean section of a bevel gearing, assuming the uniform load distribution among contact areas, by which the factor of the gearing parameters variability along the gear toothing width and the influence of manufacturing and assembly errors are ignored. It is easy to notice that obtained by [172], stresses are not even close to limiting failure ones, which makes a mistaken impression of a considerable underloading of the gearing.

Some other examples of calculations by the CONUS program are given in Appendix 5.

Thus, the developed technique and the computer program take into account the peculiarities of operation of the bevel NG in real conditions, that provides higher reliability of the calculation.

Chapter 18

SEVERAL WAYS TO INCREASE THE LOAD-BEARING CAPACITY OF NOVIKOV GEARING

The requirements applied in the recent years to market competitiveness of drives predetermined active search of reserves the load-bearing capacity increasing of Novikov gearing with unground teeth. It will be proved below that a rather effective way of such an increase was longitudinal modification of the tooth flanks.

Before considering the longitudinal modification, letus discuss the problem of the gear toothing width influence on the load-bearing capacity of NG.

18.1 INFLUENCE OF THE GEAR TOOTHING WIDTH ON THE LOAD-BEARING CAPACITY OF THE GEARING

It is known that the teeth changeover in the NG occurs only due to its axial overlap, ensured by a coefficient ε_β, which is assigned, as a rule, not less, than unity. In those cases, when a radial dimension of the gearing (gear diameters, center distance) is restricted by a drive design, there is a tendency to raise the load-bearing capacity by increasing the axial dimension, i.e. operating toothing width b_w of gears (tooth length), and, respectively, proportional to it (at other constant parameters) axial overlap ratio ε_β. The same purpose (aside from excluding the axial force component in the engagement) is often pursued in design of the herring-bone gearing, the total toothing width of half of the herring-bone in which is rather large.

If we suppose that a gearing is made without manufacturing errors, and mating parts (shafts, bearings, etc.) are absolutely rigid, then with increasing b_w (or ε_β) it is right to expect the raise of the load-bearing capacity of such a pair (which will be shortly called "ideal").

Meanwhile, as it was shown in Chapter 10, the real conditions of gearing operation, when there are errors of its manufacturing, deformations of the mating parts and so on, lead to a considerable non-uniform distribution of the transmitted load and acting stresses among contact areas, and consequently, to the essential decrease of the load-bearing capacity compared with the "ideal" pair. It is confirmed by a great number of research works and also by practice of operation of involute gears. Particularly, it is convincingly shown that, with raising of b_w (or the ratio of b_w to the pinion diameter) above a certain limit deformations of shafts deflection and torsion lead to the decreasing of the gearing load-bearing capacity [140].

As for the NG TLA, there are recommendations in bibliography about rational assignment of the value ε_β (for example, in [141]), but besides insufficient completeness (it will be described below), these recommendations are suitable, but partially, only for the "ideal" gearing, that is why their practical utility is not great.

With the appearance of methods of determination of non-uniform distribution of the load and stresses among contact areas [100, 110], and taking into account a dynamical force component in the engagement [228] (see also Chapter 10), it became possible to investigate influence of the gear toothing width on the load-bearing capacity of the NG, operating in real conditions.

Results of the given investigation are illustrated below by an example of a typical cylindrical NG with the basic rack profile according to the standard GOST 30224-96 and possessing parameters: $m=5$ *mm*, $z_1=13$, $z_2=47$, $\beta=18°$, $n_1=1500 min^{-1}$, location of gear rims of the pair – asymmetric with respect to the shaft bearing supports (that is typical of a gearing of multi-stage gearboxes). In this case, 3 versions are considered: pair *1* – an "ideal" one (to be compared), pair *2* – with surface hardened teeth, having hardness of tooth flanks not less, than *HB 570*, pair *3* – with tempered tooth flank hardness *HB 300* (pinion) and *HB 270* (gear). To ensure good visualization for pairs *2* and *3*, a not high (*9*) degree of manufacturing accuracy according to kinematics, smoothness and contact ratings was chosen, and the corresponding scale of coordinate axes for plotted diagrams of investigated parameters was chosen, where numbers *1, 2, 3* correspond to the outlined pairs.

Fig. 57 shows diagrams of the coefficient $K_{F\varepsilon}$, representing the ratio of allowable torque by bending endurance (load-bearing capacity of the gearing) with the assigned ε_β to the corresponding torque of the same gearing with $\varepsilon_\beta=1$.

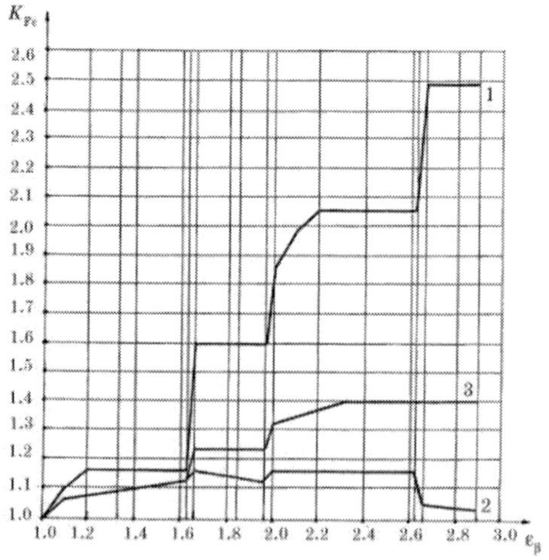

Figure 57. Diagram of dependence of $K_{F\varepsilon}$ on ε_β.

It is seen from the diagram *1* for the "ideal" pair, that there are areas where the value of $K_{F\varepsilon}$ doesn't change with raising ε_β and there are areas of its rapid increment. The type of a diagram, approximated by a polygonal line, indicates the discrete nature of engagement, where a definite integer number of theoretical contact points along the tooth length, going in and out of the engagement, participates at each phase. The ranges of ε_β, preferable for the pair *1*, are as follows: *1.0...1.2, 1.66...1.70, 2.0...2.2, 2.66...2.70* and etc. and it is appropriate to apply ranges *1.2...1.65, 1.75...1.95, 2.2...2.65, 2.75...2.95* etc. It should be noted that ranges *1.0...1.2, 2.0...2.2, 3.0...3.2* etc. are recommended in [141], but the attention was not paid to other rational areas. The analysis showed that a considerable increase of the gearing load-bearing capacity takes place at such values of ε_β, when minimum theoretical number n_{min} of contact points in all engagement phases increases. Such values of ε_β are equal to either an integer number *K*, or to a number $K+\varepsilon_q$ (Table 6). For the described example $\varepsilon_q \approx 0.66$.

(Actually, the curve $K_{F\varepsilon}$ is more smooth, with less sharp jumps, however, it doesn't make sense).

As for the gearing *2*, the pattern is quite different. If there is a definite raise of $K_{F\varepsilon}$ up to $\varepsilon_\beta=1.66$, then later its fall is observed (with a small rise only at $\varepsilon_\beta=2.0$). It is explained by the fact that standard manufacturing errors [40], deformations of shafts deflection and torsion, and reduced mass of the pair, that is, the dynamic component in the engagement [228], are increased with the raise of ε_β (i.e. b_w). That is why for the pair *2,* values of $\varepsilon_\beta>1.7$ are not desirable. When comparing pairs *1* and *2*, it is seen that with the raise of ε_β from *1.0* to *1.66* the increase of $K_{F\varepsilon}$ for the pair *1* is *1.6* times, and for the pair *2* – only by *16%*.

The pair *3* has a more favorable diagram of $K_{F\varepsilon}$, than the pair *2*, but it is also far from the diagram of the pair *1*. A positive influence of teeth surfaces run-in process, which is more intensive at their low hardness, promotes reducing of non-uniform load and stresses distribution among contact areas [129, 152] and, as a consequence, the raise of $K_{F\varepsilon}$. From diagram *3,* it is seen, that *1.66* times increase of the axial dimension (ε_β) of the pair gives *1.25* times increase of $K_{F\varepsilon}$. Besides, unlike the pair *2*, there is an area of $\varepsilon_\beta=2.0...2.3$, where also some raise of $K_{F\varepsilon}$ occurs, though *2.3* times increase of the axial dimension of the pair promotes the raise of $K_{F\varepsilon}$ by *40%*.

Comparison of the considered variants of pairs by such an important factor in gearbox industry as specific material intensity [208] (in this case expressed as the ratio of the value b_w to allowable torque at the gear), showed that, as it was expected, the pair *1* has the lowest factor, the pair *3* – the highest one, though dynamics of this factor increase is more pronounced for the pair *2* [102].

Research works showed that the diagram considered above is qualitatively repeated for other degrees of accuracy of the gearing, including its estimation by a criterion of contact endurance of the tooth flanks.

On the basis of the carried out analysis of working capacity of the NG, with non-modified tooth surfaces in the most widespread range of angles $\beta = 17^\circ ... 20^\circ$, it is possible to state the following:

1) For helical gearing with surface-hardened teeth values of $\varepsilon_\beta = 1.05...1.1$ are more preferable and the second, $\varepsilon_\beta = (1.05...1.1) + \varepsilon_q$; the application of surface-hardened teeth for herring-bone gearing, which can be considered as a helical one within the carried out research with a value b_w , equal to the total width of both halves of the herring-bone and a groove between them, is not reasonable at all.
2) For helical gearing with small tooth hardness, the order of preferable application for values ε_β is as follows: $\varepsilon_\beta = 1.05...1.1$, $\varepsilon_\beta = (1.05...1.1) + \varepsilon_q$, $\varepsilon_\beta = 2.05...2.3$, values $\varepsilon_\beta = (1.15 + \varepsilon_q)...1.95$ and $\varepsilon_\beta > 2.3$ are not recommended; for herring-bone gearing the most reasonable coefficient ε_β for half of the herring-bone is *1.05...1.15*.

18.2. Longitudinal Modification of Tooth Flanks of Cylindrical Gearing

It was shown in previous chapters that the stressed state of NG teeth depends essentially on the non-uniform stresses distribution among contact areas, which, in turn, is determined by the gearing geometry, accuracy of its manufacture, stiffness of teeth, gears, shafts, bearings, etc., and by the influence of the teeth faces.

Letus call further the period during which theoretical contact points pass the axial pitch along lines of action, as a cycle of engagement, and designate the stress concentration by a factor K_σ, representing the maximum value of the product $K_T K_b$ over the whole cycle within a certain contact area, where K_T is the part of the total transmitted torque T_Σ, acting on the given area, K_b is the coefficient, taking into account the influence of the face vicinity to the considered area (see Chapter 10).

If the initial (before the teeth run-in) operation stage of the gearing is considered and the dynamic force component in the engagement [228], insignificant at rotational speeds less than *10 m/s*, is excluded, then at the given tooth geometry, the acting stress can be represented as a function of the product

$$\sigma = \sigma(K_\sigma \cdot T_\Sigma). \tag{18.1}$$

Reduction of K_σ and, consequently, the increase of T_Σ (that is, load-bearing capacity) at the given stress σ can be obtained in several ways. A conventional method is to increase the accuracy of gears manufacturing and assembly, which, however, is restricted by increasing the cost of the design unit. Another method is a rational design of the gearing unit, influencing the rigidity of the whole design unit, but this method has a limited effectiveness.

Significant results, as it will be shown below, are achieved by a rather simple, third method – longitudinal localization of tooth contacting surfaces, which leads not only to reducing the negative influence of faces, but also, which is important, to equalization of stresses among contact areas.

In the correct statement, a similar task is reduced to finding a certain three-dimensional curved tooth line, at which minimum K_σ is obtained. However, here there are difficulties, not of mathematical but of manufacturing character, connected with possibility of practical implementation of the synthesized geometry. That is why it is reasonable to apply the known available manufacturing methods of modification, which donot lead to a considerable rise in the cost of the tooth cutting process, and to make the choice of modification parameters, basing on the achievement of maximum load-bearing capacity of the gearing.

The known approaches of longitudinal localization can be reduced to the following ones:

a) cutting of one element of a gear pair with the corrected (comparing with the nominal one) helix angle;
b) longitudinal tooth surfaces flanking at faces;
c) giving to the tooth surfaces a slightly convex (barrel-shaped) required curvature shape along the length.

Strictly speaking, the first approach as a method that does not cause the change (modification) of tooth surfaces, can be related to modification methods only conditionally and is considered only due to its certain effectiveness; along with the second approach, it is applied mainly to the cylindrical gearing. The third approach is applied both to cylindrical and bevel gearing and shows high effectiveness [118, 131, 229].

At first sight it may seem that it is sufficient to follow some well-known recommendations, for example, on longitudinal tooth flanking of the NG [172] or on obtaining barrel-shaped teeth of the quasi-involute gearing [127], and the problem will be solved. Indeed it is far from the truth. The idea is not only in a formal application of some or other modification method, but in determining such modification parameters, which ensure a visible effect. According to the research works, application of unreasonable (sometimes chosen randomly) modification parameters (flank size, arrow of convexity of a barrel-shaped tooth, etc.) often doesn't give the required effect, but what is more, it even leads to worsening of the gearing working capacity, which is sometimes considerable [121].

In our opinion, study of the longitudinal modification influence on the load-bearing capacity and the setting of the rational modification parameters is possible, only with application of the developed method of determination of non-uniform load and stresses distribution among contact areas with account of boundary effects, described in Chapter 10.

Without detriment to final conclusions, letus consider one of the most important factors – accuracy of gears manufacturing and assembly – as a factor influencing the non-uniform load and stresses distribution among contact areas. In this case, the system of equations (10.10), describing the elastic equilibrium and consistency of displacements of contacting teeth, takes the form [118]:

$$\begin{cases} (K_{b1}T_1)^x + d_1 = (K_{b2}T_2)^x + d_2 = \ldots = (K_{bn}T_n)^x + d_n; \\ \sum_{i=1}^{n} T_i = T_\Sigma, \end{cases} \tag{18.2}$$

where n is the number of theoretical contact points (areas) in the engagement, determined by the degree of axial overlap of teeth;

T_i is the torque, acting on the "*i*"-th contact area; $d_i = (\Delta_i + \delta f_i + \delta l_i)/A$; x, A are constants, depending on the geometry of the basic rack profile and the gearing [163]; Δ_i is the kinematic clearance between contact points, appearing due to manufacturing and assembly errors of the pair, estimated in a stochastic aspect with a given "percent of risk" [100, 129]; $\delta f_i, \delta l_i$ is the kinematic clearance, appearing in the engagement for longitudinal flanking and for the manufacture of a barrel-shaped tooth correspondingly, its selection makes it possible to influence the redistribution of loads and stresses.

If there are the noted clearances, then the number of roots p of the system (18.2) solution may become less, than *n*, that is, $p \le n$. Taking into account that $K_{bi} > 0$ and roots T_i of the system must also be positive, it is convenient to solve the system by a consequent introduction of equations, satisfying the initial condition of $(K_{b1}T_1)^x > (K_{bi}T_i)^x + (d_i - d_1)$.

The result of the system (18.2) solution is the determination of coefficients $K_{Ti} = T_i / T_\Sigma$, the number of which (and also the number of actually contacting areas) is equal to p.

Coefficients of bending $(K_{F\sigma i})$ and contact $(K_{H\sigma i})$ stresses on contact areas are determined on the basis of K_{Ti}:

$$K_{F\sigma i} = K_{Ti} K_{Fbi}, \tag{18.3}$$

$$K_{H\sigma i} = K_{Ti} K_{Hbi}. \tag{18.4}$$

The value of K_{Fb} is changed according to a relationship, close to an exponential one and lies within the range from *1* to *2* [100], and the value K_{Hb} changes from *1* to the value, determined by the helix angle β, the ratio of main reduced curvatures of contacting surfaces and the ratio of distance of an elliptical area center from the face to the value of longitudinal semi-axis of the ellipse [110].

The engagement cycle is divided into rather small steps. At each step, the set of coefficients $K_{F\sigma i}$, $K_{H\sigma i}$ is determined, according to which maximum values $K_{F\sigma}$, $K_{H\sigma}$ are found, which are the design values in analysis of the tooth stressed state.

Numerical simulation for a wide range of geometrical parameters of gears is carried out by means of the developed computer program MODIF. Here it should be noted that the solution of the system (18.2) does not have restrictions both on the number of theoretical contact points *n* and on the number of steps of the engagement cycle division.

Coefficients K_σ were determined for the gearing, manufactured with high-hardened (*HB 570)* tooth surfaces and with thermally improved (*HB 270)* teeth. Since the limiting failure criterion for high-hardened teeth is, as a rule, bending endurance of teeth, then the coefficient $K_{F\sigma}$ was determined for them; for the thermally improved gearing, where the limiting criterion is the contact endurance of tooth surfaces, the coefficient $K_{H\sigma}$ was determined. Taking into account that the coefficient K_σ depends, in addition, on the value T_Σ of the load, the latter was accepted by the allowable torque according to corresponding limiting criterion for the gearing of the 8th degree of accuracy by smoothness and contact ratings [40], for which the stress (18.1) becomes equal to the allowable one.

It was found out that the coefficient K_σ is discretely changed depending on the minimum number of theoretical contact points in the cycle, and the latter depends on the range, where the axial overlap ratio ε_β is located, and on values of the phase overlap ratio ε_q (Table 6).

Table 30 contains geometrical parameters and design loads for some typical calculated pairs, based on the basic rack profile according to the standard GOST 30224-96, for which the obtained results will be shown below.

Table 30. Parameters of calculated pairs.

Pairs N	m, mm	z_1	z_2	β,°	ε_q	b_w, mm	ε_β	T_Σ at the driven gear, N·m	
								Tooth hardness HB 570	Tooth hardness HB 270
1a	3.15	15	45	19.3	0.651	33	1.102	1100	250
1b						51	1.703	1250	300
2a	5	15	45	20.3	0.645	50	1.104	5000	1200
2b						76	1.679	6000	1500
3a	8	15	44	19.5	0.650	83	1.102	20000	6000
3b						126	1.674	23000	7500

Pairs of the group "a" (1a, 2a, 3a) have the axial overlap ratio ε_β within the range $1 \leq \varepsilon_\beta < 1+\varepsilon_q$, and pairs of the group "b" (1b, 2b, 3b) – within the range $1+\varepsilon_q \leq \varepsilon_\beta < 2$. A minimum number of theoretical contact points in the cycle for pairs of the group "a" is equal to 2, and for pairs of the group "b" it is equal to 3.

Letus consider a cutting method of one element of the pair with the corrected helix angle.

The given method is reasonable only in small-batch and piece production of gearing, for which it is possible to measure directly deviations of teeth directions of the pinion $(F_{\beta r1})$, of the gear $(F_{\beta r2})$ and misalignment f_{yr} of axes of the pair (by housings of the casing), which relate to the contact ratings [40]. In this case, the radial clearance Δ_i involved into the system (18.2) will consist of two components, one of which takes into account errors, related to the smoothness rating in a stochastic aspect, and another one – the algebraic sum of values $F_{\beta r1}$, $F_{\beta r2}$ and f_{yr}.

Fig. 58*a* conditionally shows a pair of teeth of basic racks for a pinion 1 and a gear 2, which don't have errors by the contact rating ($\delta a = 0$) and contacting with each other along a straight line, and Fig. 58*b* shows the same pair with errors by the contact ratings, leading to kinematic clearance $\delta a \neq 0$ at one of the gear rim faces. Since errors $F_{\beta r}$ and f_{yr} have a practically linear character along the tooth length, their negative action may be compensated by the helix angle β correction by the value $\Delta\beta = \pm arccot(\delta a / b_w)$ during teeth cutting of one of the pair elements (for example, pinion).

In the most unfavorable case, the noted deviations will be equal to limiting values corresponding to the given degree of accuracy, directed to one side, and then, taking into account the regulated [40] relations between these deviations, $\delta a = \delta a_{max} = F_{\beta 1} + F_{\beta 2} + f_y = 2.5F_\beta$.

However, although such case is possible, it is hardly probable. That is why it seemed expedient to additionally consider the case with the most probable value of δa, which, according to our calculations, is equal to $\delta a = \delta a_v = 1.37F_\beta$.

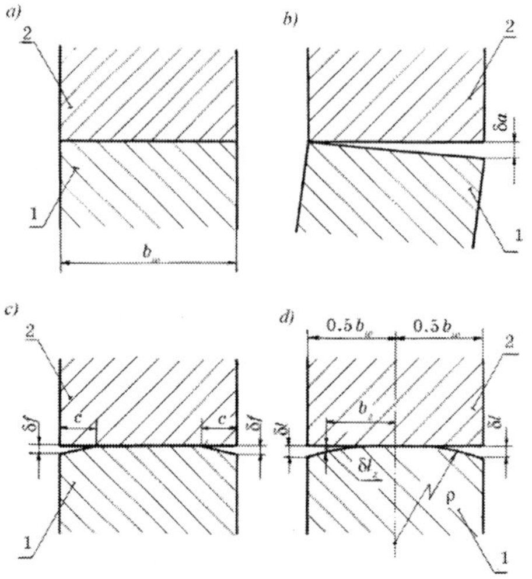

Figure 58. Scheme of contacting of non-modified teeth: *a*) without manufacturing errors; *b*) with errors according to contact ratings. Scheme of contacting of modified teeth (of the pinion 1): *c*) with longitudinal flanking; *d*) barrel-shaped.

Table 31 shows the results of calculation of coefficients $K_{F\sigma}$ and $K_{H\sigma}$ for the gearing 3a, 3b at different values of δa (degrees of accuracy by smoothness and contact ratings are equal).

Letus consider further a method of longitudinal tooth flanking.

In this method, for one element of the pair (pinion is more convenient) equal flanks near both faces with the depth δf and length c are removed, as it is shown in Fig. 58*c*. There are different methods of longitudinal tooth flanking in NG [87], for which the line of flank is a straight line or close to it.

A question of longitudinal flanking has so far been considered only from the point of view of improving the dynamic characteristics of the engagement by preventing the face edge impacts [87, 172]. That is why there are recommendations in which it is suggested either to copy parameters of longitudinal flanking of involute gearing [87], that is principally wrong, or to implement one-side flanking (only at the entrance in the engagement) to a considerable depth, keeping the non-flanked part of the tooth length not less, than the axial pitch [172]. In such an approach, the face edge is completely excluded from the operation, but the main cycle of engagement (at the non-flanked part) remains the same, and equalization of stresses among contact areas does not occur. Moreover, according to calculations in a number of cases, deep flanking at a small length, unrelated to the gearing accuracy, leads even to increasing of K_σ and decreasing of the load-bearing capacity. The authors of this approach have reasons to recommend finally to choose the flanking parameters experimentally in each particular case [172].

Table 31. Values of coefficients $K_{F\sigma}, K_{H\sigma}$ for the gearing 3a, 3b.

Pair N	δa	Values of $K_{F\sigma}$						Values of $K_{H\sigma}$					
		Degree of accuracy						Degree of accuracy					
		7	8	9	10	11	12	7	8	9	10	11	12
3a	δa_{max}	0.81	0.97	1.21	1.50	1.89	2.00	0.91	0.99	1.17	1.26	1.26	1.26
	δa_v	0.74	0.84	1.01	1.25	1.52	1.89	0.85	0.93	1.05	1.18	1.26	1.26
	0	0.71	0.74	0.81	0.95	1.18	1.49	0.80	0.85	0.93	1.03	1.18	1.18
3b	δa_{max}	0.66	0.81	1.01	1.27	1.56	1.95	0.78	0.91	1.02	1.18	1.26	1.26
	δa_v	0.60	0.68	0.83	1.09	1.39	1.68	0.72	0.84	0.96	1.11	1.18	1.26
	0	0.53	0.56	0.62	0.76	0.99	1.33	0.63	0.70	0.81	0.97	1.15	1.18

Our research works showed that tooth flanking in NG should be carried out not only to exclude close to faces parts from operation, but also to lessen their stressed state and reduce the non-uniform stresses distribution among all areas. Parameters of such flanking, of course, will depend on ε_β and the gearing accuracy. Here the flanking must be two-sided, as it is shown in Fig. 58*c*, because the direction of errors vector is unknown beforehand, and any face may appear to be overstressed. On the other hand, as it was already noted above, at mean tangential speeds the dynamical force component in the engagement is not large, and its further reduction should be expected at any flanking depth, combined with equalization of the load distribution among contact areas.

Letus represent the relations which we obtained for determination of longitudinal flanking parameters (in *mm*), application of which gives the best results.

For gearing of the group "a" with high-hardened teeth

$$\delta f = 6.08 \cdot 10^{-6} k^{3.94} \text{ - for } k = 7...11, \tag{18.5}$$

$$\delta f_{k=12} = 2.5 \delta f_{k=11},$$

with thermally improved teeth – the same relationship (18.5), but for $k = 9, 10,$ here

$$\delta f_{k=11} = 2\delta f_{k=10}, \; \delta f_{k=12} = 2\delta f_{k=11} = 4\delta f_{k=10}. \quad (18.6)$$

For gearing of the group “b” at any tooth hardness

$$\delta f = 8.47 \cdot 10^{-9} k^{6.85}, \quad (18.7)$$

for high-hardened teeth the formula (18.7) being applicable at $k = 8...12,$ and for thermally improved ones – at $k=9...12.$

The flank length for all considered types of gearing can be accepted equal to

$$c=2m. \quad (18.8)$$

The letter k here designates the number of the so-called reduced degree of accuracy:

$$k = 0.65k_p + 0.35k_k. \quad (18.9)$$

The formula (18.9) reflects the fact, established by calculations, of greater effect of deviations by the smoothness rating effect on K_σ than deviation by the contact rating.

From (18.5) – (18.7) it is seen that the flank depth δf does not depend on the gearing module, but depends, as it was expected, on ε_β and k.

For degrees of accuracy, not included in the relationships (18.5) – (18.7), the longitudinal flanking effect is less than 3%, that is why results for them are not shown.

Table 32 shows results of calculation of coefficients $K_{F\sigma}$ and $K_{H\sigma}$ for flanked gearing 1a, 1b with application of parameters δf, c, determined according to (18.5) – (18.8), and similar coefficients $K^0_{F\sigma}$ and $K^0_{H\sigma}$ for non-flanked ($\delta f = 0$) teeth. (Here and further, the upper index “0” of stress concentration factors means that these factors refer to non-modified teeth).

As it is seen from Table 32, flanking in terms of strength “transfers” the gearing to a higher degree of accuracy. For example, the gearing 1a of the 10th degree of accuracy has $K_{F\sigma}$ at the level of value $K^0_{F\sigma}$ for a gearing of the 8th degree, the gearing 1a and 1b of the 12th degree of accuracy have $K_{H\sigma}$ less than $K^0_{H\sigma}$ for the corresponding gearing of the 9th degree, etc.

Letus consider, finally, the modification method, consisting in forming a convex (barrel-like) surface of the tooth.

This modification method is successfully applied for involute gearing. For cylindrical NG it, as far as we know, is proposed by us for the first time [109].

Table 32. Values of coefficients $K^0_{F\sigma}, K_{F\sigma}$ and $K^0_{H\sigma}, K_{H\sigma}$ for the gearing 1a, 1b.

Pair N	$K^0_{F\sigma}$, $K_{F\sigma}$	Degree of accuracy k						$K^0_{H\sigma}$ $K_{H\sigma}$	Degree of accuracy k			
		7	8	9	10	11	12		9	10	11	12
1a	$K^0_{F\sigma}$	0.88	1.05	1.31	1.58	1.89	2.00	$K^0_{H\sigma}$	1.21	1.22	1.22	1.22
	$K_{F\sigma}$	0.77	0.85	0.96	1.02	1.14	1.14	$K_{H\sigma}$	1.00	1.03	1.00	1.00
1b	$K^0_{F\sigma}$	-	0.87	1.15	1.48	1.76	1.94	$K^0_{H\sigma}$	1.15	1.21	1.22	1.22
	$K_{F\sigma}$	-	0.71	0.87	0.98	1.06	1.08	$K_{H\sigma}$	1.00	1.00	1.00	1.00

The method is illustrated in Fig. 58*d*, where surface of the element *1* (pinion) in longitudinal direction has a slightly convex (barrel-like) shape, symmetrical with respect to the middle of the gear rim, with maximum ease-off at the faces (arrow of convexity) δl and curvature radius ρ in the middle point of the toothing width. Such a shape can be obtained, for example, by a known from the practice of involute gearing by diagonal hobbing with variable tooth thickness of the hob [247], or by shaving on pendulum-table machine-tools with help of shaving cutters having Novikov teeth [207]. Compared with tooth milling, tooth shaving is preferable, it has an advantage that, simultaneously with the surface modification, its condition is improved and some deviations by smoothness rating are reduced [207].

Because of the small δl value compared with the ρ value, the plane section with the longitudinal shape of the rack tooth can be approximated with a sufficient accuracy by a circumference arc of radius ρ, the current value δl_z of the ease-off at the tooth length b_z being determined by a parabolic relation [127]: $\delta l_z = b_z^2/(2\rho) = 4\delta l(b_z/b_w)^2$.

In research works, it is established that the parameter δl can be accepted according to the same relationships (18.5) – (18.7), as the flank depth δf.

Table 33 contains results of calculation of coefficients $K_{F\sigma}$ and $K_{H\sigma}$ for the gearing 2a, 2b for values $\delta l = \delta f$ (18.5) – (18.7) and corresponding coefficients $K^0_{F\sigma}$ and $K^0_{H\sigma}$ for $\delta l = 0$.

It is seen from Table 33 that, as in the case of flanking (Table 32), the gearing with barrel-shaped teeth of rough degrees of accuracy have a stress concentration at the level of the gearing, manufactured more precisely by *1 - 3* degrees, than the gearing with non-modified teeth.

On the basis of the stated above, it is possible to evaluate the effect of the stress reduction for the tooth modification by comparing concentration factors of modified and non-modified teeth.

Letus designate: $\theta_{F\sigma m}(\theta_{H\sigma m})$ and $\theta_{F\sigma v}(\theta_{H\sigma v})$ are ratios of stress concentration factors for $\delta a = \delta a_{max}$ and $\delta a = \delta a_v$, respectively, to concentration factors for $\delta a = 0$ (Table 31); $\theta_{F\sigma f}(\theta_{H\sigma f})$ and $\theta_{F\sigma l}(\theta_{H\sigma l})$ are ratios of concentration factors $K^0_{F\sigma}$ $(K^0_{H\sigma})$ for non-modified teeth ($\delta f = 0$ and $\delta l = 0$) to corresponding concentration factors $K_{F\sigma}$ ($K_{H\sigma}$) of flanked ($\delta f \neq 0$) and barrel-shaped ($\delta l \neq 0$) teeth with the determined according to (18.5) – (18.8) modification parameters (Tables 32, 33).

Table 33. Value of coefficients $K^0_{F\sigma}, K_{F\sigma}$ and $K^0_{H\sigma}, K_{H\sigma}$ for the gearing 2a, 2b.

Pair N	$K^0_{F\sigma}$, $K_{F\sigma}$	Degree of accuracy k						, $K_{H\sigma}$	Degree of accuracy k			
		7	8	9	10	11	12		9	10	11	12
2a	$K^0_{F\sigma}$	0.75	0.84	1.01	1.26	1.54	1.87	$K^0_{H\sigma}$	1.16	1.23	1.29	1.29
	$K_{F\sigma}$	0.71	0.76	0.85	0.95	1.02	1.06	$K_{H\sigma}$	1.00	1.00	1.00	1.00
2b	$K^0_{F\sigma}$	-	0.64	0.81	1.06	1.40	1.70	$K^0_{H\sigma}$	1.06	1.21	1.24	1.29
	$K_{F\sigma}$	-	0.59	0.68	0.76	0.84	0.95	$K_{H\sigma}$	0.95	1.00	1.00	1.00

The summary Table 34 contains the noted ratios for the considered pairs, which are the indicators of effectiveness of bending and contact stresses reduction (dashes in Table 34 mean that the effect of stress reduction is less than *3%*).

It is seen from Table 34 that the effect by bending strength, in almost all cases, considerably exceeds the effect by contact strength. The effect from modification of flanked and barrel-shaped teeth is approximately the same iIt is significantly increased with the decrease of the gearing manufacture accuracy and slightly reduced with its module raising, remaining at the comparable level for the gearing from groups "a" and "b". When applying a method of correction of the angle β, the greatest effect is achieved for the 9th, 10th degrees of accuracy, reducing further with the accuracy decreasing. It is explained by the fact that, for rough (*11, 12*) degrees of accuracy deviations by the smoothness rating, approaching considerable values and not compensated by this method, begin to dominate (negatively influencing) in distribution of load and stresses among contact areas. However, it is important that, for the most widespread in machine-building industry degrees of accuracy (*8...10*) of gearing manufacturing with unground teeth for all considered modification methods, the effect of stress reduction, as it is seen from Table 34, is significant.

The factors $\theta_{F\sigma}, \theta_{H\sigma}$ will be more increased with taking into account deformations of mating parts (shafts, bearings, etc.), but in a certain degree it will be compensated by run-in of teeth surfaces, at which coefficients $K_{F\sigma}, K_{H\sigma}$ are decreased insignificantly for high-hardened teeth and more considerably for teeth with low hardness [152].

Letus note another important circumstance.

Table 34. Results of the tooth modification effectiveness.

Pairs*N*		Degree of accuracy *k*							Degree of accuracy *k*					
		7	8	9	10	11	12		7	8	9	10	11	12
1a	$\theta_{F\sigma m}$	1.41	1.50	1.48	1.34	1.13	1.13	$\theta_{H\sigma m}$	1.22	1.07	1.07	1.07	-	-
	$\theta_{F\sigma v}$	1.21	1.27	1.26	1.18	1.13	1.13	$\theta_{H\sigma v}$	1.14	1.07	1.07	1.07	-	-
	$\theta_{F\sigma f}$	1.14	1.24	1.36	1.55	1.66	1.75	$\theta_{H\sigma f}$	-	-	1.21	1.18	1.22	1.22
	$\theta_{F\sigma l}$	1.12	1.20	1.34	1.58	1.71	1.82	$\theta_{H\sigma l}$	-	-	1.21	1.20	1.22	1.22
1b	$\theta_{F\sigma m}$	1.52	1.58	1.49	1.36	1.16	1.13	$\theta_{H\sigma m}$	1.18	1.16	1.07	1.07	1.07	-
	$\theta_{F\sigma v}$	1.28	1.36	1.37	1.25	1.05	1.13	$\theta_{H\sigma v}$	1.11	1.09	1.06	1.07	1.07	-
	$\theta_{F\sigma f}$	-	1.23	1.32	1.51	1.66	1.80	$\theta_{H\sigma f}$	-	-	1.15	1.21	1.22	1.22
	$\theta_{F\sigma l}$	-	1.22	1.43	1.68	1.79	1.94	$\theta_{H\sigma l}$	-	-	1.14	1.21	1.22	1.22
2a	$\theta_{F\sigma m}$	1.22	1.41	1.50	1.54	1.40	1.14	$\theta_{H\sigma m}$	1.17	1.27	1.16	1.07	1.07	-
	$\theta_{F\sigma v}$	1.08	1.20	1.29	1.30	1.24	1.14	$\theta_{H\sigma v}$	1.07	1.14	1.09	1.07	1.07	-
	$\theta_{F\sigma f}$	1.05	1.13	1.23	1.34	1.54	1.75	$\theta_{H\sigma f}$	-	-	1.16	1.23	1.29	1.29
	$\theta_{F\sigma l}$	1.06	1.11	1.19	1.33	1.51	1.76	$\theta_{H\sigma l}$	-	-	1.16	1.23	1.29	1.29

Table 34. (Continued)

Pairs$N$		Degree of accuracy k							Degree of accuracy k					
		7	8	9	10	11	12		7	8	9	10	11	12
2b	$\theta_{F\sigma m}$	1.34	1.54	1.64	1.59	1.45	1.23	$\theta_{H\sigma m}$	1.27	1.25	1.22	1.07	1.06	1.06
	$\theta_{F\sigma v}$	1.17	1.29	1.39	1.43	1.32	1.12	$\theta_{H\sigma v}$	1.18	1.16	1.18	-	1.06	1.06
	$\theta_{F\sigma f}$	-	1.09	1.18	1.32	1.48	1.65	$\theta_{H\sigma f}$	-	-	1.06	1.21	1.24	1.29
	$\theta_{F\sigma l}$	-	1.08	1.19	1.39	1.67	1.79	$\theta_{H\sigma l}$	-	-	1.12	1.21	1.24	1.29
3a	$\theta_{F\sigma m}$	1.14	1.31	1.49	1.58	1.60	1.34	$\theta_{H\sigma m}$	1.14	1.16	1.26	1.22	1.07	1.07
	$\theta_{F\sigma v}$	1.04	1.14	1.25	1.32	1.29	1.27	$\theta_{H\sigma v}$	1.06	1.09	1.13	1.15	1.07	1.07
	$\theta_{F\sigma f}$	1.03	1.06	1.15	1.24	1.36	1.58	$\theta_{H\sigma f}$	-	-	-	1.08	1.21	1.26
	$\theta_{F\sigma l}$	1.05	1.05	1.13	1.23	1.34	1.59	$\theta_{H\sigma l}$	-	-	-	1.08	1.21	1.26
3b	$\theta_{F\sigma m}$	1.25	1.45	1.63	1.67	1.58	1.47	$\theta_{H\sigma m}$	1.24	1.30	1.26	1.22	1.10	1.07
	$\theta_{F\sigma v}$	1.13	1.21	1.34	1.43	1.40	1.26	$\theta_{H\sigma v}$	1.14	1.20	1.19	1.14	-	1.07
	$\theta_{F\sigma f}$	-	1.08	1.10	1.22	1.36	1.51	$\theta_{H\sigma f}$	-	-	-	-	1.17	1.22
	$\theta_{F\sigma l}$	-	1.08	1.10	1.24	1.50	1.69	$\theta_{H\sigma l}$	-	-	1.05	1.09	1.18	1.22

Since the coefficient K_σ depends non-linearly on the load (see Chapter 10) – with increasing of the load the value K_σ is reduced, but in a lesser degree, and vice versa – the factor of effectiveness of the load-bearing capacity increase, expressed by the ratio of allowable load for modified teeth to such load for non-modified ones (that is, under fixed allowable stress, regulated by standards), will always be greater than the stress reduction effectiveness under the given load T_Σ, estimated by the factor θ_σ (Table 34).

Research works and calculations showed that the effectiveness of load-bearing capacity increase for the cylindrical NG, due to longitudinal modification of teeth surfaces, can approach *2* and more according to bending and *(40...45)%* of contact endurance.

Of course, such a considerable increase of the gearing load-bearing capacity can not always be practically implemented due to different reasons, for example, because of the limited strength of shafts, bearings lifetime, etc. Nevertheless, the given information describes the revealed significant reserve of increasing the energy intensity and reducing the material intensity of the NG due to longitudinal modification of its tooth surfaces.

Hence, the main conclusion can be done as follows;

Longitudinal modification of tooth surfaces, carried out with the recommended above modification parameters, is a technologically simple, and at the same time effective, way for reduction of the tooth stress level and material intensity of the drive, and also for the increase of the load-bearing capacity (up to *2* times and more) and competitiveness of the cylindrical NG in the world market.

18.3. Longitudinal Modification of Tooth Flanks of Bevel Gearing

It is known that during cutting of both straight [122, 195] and circular [148, 214] teeth of bevel gearing, based on rectilinear basic rack profile (quasi-involute gearing), longitudinal ease-off of the contacting tooth surfaces (longitudinal modification) is widely applied in order to prevent operation of face tooth edges and premature gearing failure.

The analysis of the bevel NG TLA operation, described in Chapter 17, showed that if there are manufacturing and assembly errors, then contacting of conjugated teeth occurs with edge impacts, and distribution of the transmitted load between the contact areas is extremely non-uniform. Due to it, longitudinal modification of the bevel NG TLA as a method of influence on the noted negative factors is quite relevant.

Research of a problem of longitudinal modification influence on the character of operating load distribution can be carried out by considering the system (17.17) with all parameters included in it. Here, special attention should be paid to the determination of the angular clearance value, caused by the longitudinal modification.

Designating mean, external and current cone distances by R, R_e, R_z, correspondingly, letus write the following approximate expression for a current angular clearance δl_z, caused by longitudinal (barrel-shaped) modification:

$$\delta l_z = (R_z - R_b)^2 \delta l_T / [r_{2z}(R_e - R)^2], \tag{18.10}$$

where δl_T is the given (symmetrical at the faces) linear deviation of actual nominal tooth line from theoretical one in a face plane; $R_b = R + b_0 b_w$, b_0 is the longitudinal coordinate of modification center (theoretical touch point of contacting surfaces of two generating gears), divided by the tooth length b_w.

Clearances δl_z, algebraically added to clearances $\overline{\Delta}_z$, form a new pattern of load distribution and, consequently, SSS of teeth on contact areas at different phases of engagement. Assigning values $\delta l_T, b_0$, it is possible to reduce maximum bending and contact stresses and increase the load-bearing capacity of the gearing.

In order to create recommendations on assigning δl_T, we carried out simulation (by means of the computer program CONUS) of an iterative process of load-bearing capacity determination of certain orthogonal bevel NG TLA with different tooth hardness, with the basic rack profiles according to standards GOST 15023-76 and GOST 30224-96 in the following range of geometrical parameters: mean longitudinal overlap ratio mean helix angle $\beta_n = 25^\circ ... 45^\circ$, gear ratio of the pair $u = 1...6.3$, tool module $m_0 = 3.15...10\ mm$, degrees of accuracy by smoothness and contact ratings *8...10* according to standards GOST 1758-81 and GOST 9368-81, coefficient of gear toothing width $K_{be} = b_w / Re = 0.25$, parameter $d_{0R} = d_0 / (R \sin \beta_n) = 3$, where d_0 is the nominal diameter of the cutter head, shift factors $x^*_{n1} = 0, x^*_{\tau 1} = 0$, $b_0 = 0$, "risk level" for calculation of probable clearances is *5%*. The gears were accepted as being manufactured with axial tooth shape III [38], allowable stresses by bending and contact endurance were accepted for high-hardened teeth *650 MPa* and *1200 MPa*, respectively, for thermally improved teeth – *320 MPa* and *600 MPa*, respectively. Structural design is the most widespread: the pinion-shaft and the gear are double-seat (on tapered roller bearings), the pinion is located in cantilever, made with left tooth inclination, having a concave flank.

Taking into account great variability of the influencing parameters, the design of the calculated bevel pairs was based on a principle of assigning maximum possible module m_n, that, as it was noted, is favorable for both bending and contact endurance of NG teeth. In this case, a minimum number of pinion teeth $z_{min} = 6$ and maximum mean helix angle $\beta_{n\,max} = 45^\circ$ were the limiting factors. In accordance with the noted principle and taking into account the described above accepted parameters, the number z_c of generating gear teeth was determined as $z_c = C / \tan \beta_n$, where C depends only on $\varepsilon_{\beta n}$. The ratio $t = m_n / m_0$ was determined as $t = max\{m_n / m_{ne}, m_n / m_{ni}\}$, where m_{ne}, m_{ni} are normal modules on external and internal faces of the gearing, respectively.

Table 35 shows the main geometrical parameters of the calculated pairs.

Taking into account the rather complicated character of influence of the gearing geometrical parameters on δl_T, it became possible to obtain an approximating (with accuracy up to *4%*) dependence of δl_T only from the module m_0 [130, 131]:

$$\delta l_T = 10^{-3} \mu A_0 \cdot m_0^b, \tag{18.11}$$

where A_0, b are chosen from the Table 36 [131].

Table 35. Main geometrical parameters of calculated pairs.

Parameter	Value								
$\varepsilon_{\beta n}$	1.25			1.75			2.25		
C	27.49			38.49			49.48		
β_n, °	25	35	45	25	35	45	25	35	45
z_C	58.95	39.26	27.49	82.53	54.96	38.49	106.11	70.67	49.48
t	1.092	1.096	1.124	1.092	1.096	1.124	1.092	1.096	1.124
d_0 / m_0	45.02	45.18	46.34	63.02	63.26	64.87	81.03	81.33	83.41
R / m_n	32.52	23.96	19.44	45.53	33.55	27.21	58.54	43.13	34.99
b_w / m_n	9.292	6.846	5.554	13.01	9.585	7.775	16.73	12.32	9.996

For the gearing with the basic rack profile according to the standard GOST 15023-76 $\mu = 1$ (high-hardened teeth) and $1.05...1.1$ (thermally improved teeth), and for the gearing with the basic rack profile according to the standard GOST 30224-96 $\mu = 1.2...1.3$ regardless of the tooth hardness.

Intermediate values can be obtained by linear interpolation. For example, if it is required to determine δl_T for a gearing of the 9th degree of accuracy with $\varepsilon_{\beta n} = 1.25$, m_0=4mm, u=3, $\beta_n = 40°$, 4 values should be calculated: for $\beta_n = 35°$, u=2 - δl_T=0.061 mm, for $\beta_n = 35°$, u=4 - δl_T=0.064 mm, for β_n=45°, u=2 - δl_T=0.083 mm, and for $\beta_n = 45°$, u=4 - δl_T=0.082 mm. Then the required value δl_T =0.25(0.061+0.064+0.083+0.082)=0.0725 mm.

According to research works, a diagram of relation of the allowable load T_P from δl_T in the range of optimal values of δl_T (i.e. maximum values of T_P) has a slightly inclined character (Fig. 59), that is why the coefficient μ in the formula (18.11) changes within a certain range. Values of T_P are decreasing much more abruptly beyond the optimal area.

The analysis of calculation results showed that by means of a rational choice of δl_T, it is possible to achieve a considerable increase of the load-bearing capacity of the gearing. In the process of simulation, it was determined that the ratio $\kappa = T_P / T_0$ of the allowable transmitted load T_P for a rationally chosen value of δl_T to allowable load T_0 of the same gearing without modification (δl_T=0) ranges from 1.2 to 1.8 for the 8th degree and κ>2 for the 10th degree of accuracy. Such a considerable increase of the load-bearing capacity of the gearing is achieved mainly due to excluding from operation of more stressed tooth parts, close to the faces, and equalization of the transmitted load among contact areas. It is determined that, for the gearing with high-hardened teeth, with angles $\beta_n < 30°$ and gear ratios $u < 4$, a limiting criterion of its working capacity is the bending endurance of teeth, and in other cases – the contact endurance of tooth flanks. For the gearing with thermally improved teeth, the contact endurance is the limiting factor practically in all cases.

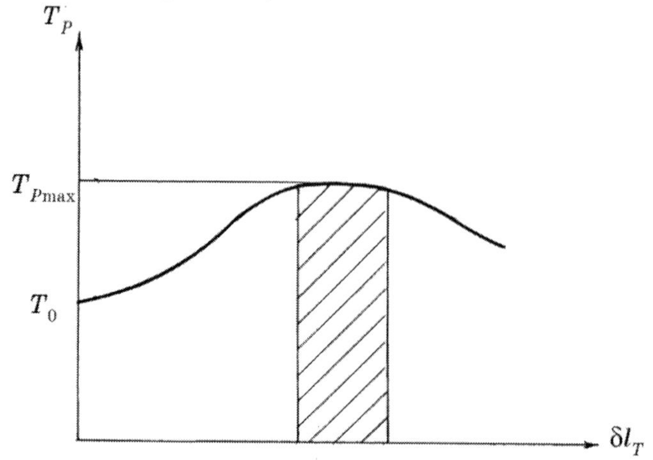

Figure 59. Typical diagram $T_P = f(\delta l_T)$: T_P – allowable torque for modified teeth; T_0 – allowable torque for non-modified teeth; δl_T – face ease-off of modified tooth surface (the area of optimal values δl_T is shaded).

On the basis of the stated above, the following conclusions can be made:

1) The recommended relation (18.11), combined with Table 36, allows to determine rather easily the rational values of tooth longitudinal modification parameters in engineering practice in a wide range of geometrical characteristics of bevel NG.
2) Taking into account design and operation peculiarities of bevel NG TLA, it should be applied only with longitudinal tooth modification, which is a way of considerable increase of the gearing load-bearing capacity (see Appendix 5), the effect increasing with reduction of its manufacturing accuracy.

Table 36. Coefficients A_0, *b*.

$\varepsilon_{\beta n} = 1.15 ... 1.35$

$\beta_n = 25°$

Degree of accuracy according to smoothness rating	$u=1$		$u=2$		$u=3$		$u=4$	
	A_0	b	A_0	b	A_0	b	A_0	b
8	3.773	0.600	4.095	0.467	7.910	0.872	17.812	0.756
9	9.288	0.232	10.582	0.133	9.684	0.849	16.700	0.883
10	24.018	-0.038	26.928	-0.121	23.635	0.509	27.720	0.701
$\beta_n = 35°$								
8	24.747	0.387	20.470	0.683	18.054	0.782	22.219	0.717
9	24.001	0.475	23.920	0.673	22.912	0.740	21.268	0.796
10	39.886	0.280	28.288	0.644	24.461	0.793	25.093	0.804

Table 36. (Continued)

Degree of accuracy according to smoothness rating	u=1		u=2		u=3		u=4	
	A_0	b	A_0	b	A_0	b	A_0	b
$\varepsilon_{\beta n} = 1.15...1.35$								
$\beta_n = 45^\circ$								
8	29.313	0.642	23.056	0.819	22.732	0.831	-	-
9	38.283	0.556	28.687	0.765	27.195	0.793	-	-
10	43.558	0.554	35.426	0.721	33.998	0.747	-	-
$\varepsilon_{\beta n} = 1.65...1.85$								
$\beta_n = 25^\circ$								
8	4.781	0.420	5.686	0.276	6.371	0.664	17.603	1.101
9	17.416	-0.116	16.304	-0.133	13.006	0.299	10.996	1.249
10	50.532	-0.351	46.399	-0.324	60.630	-0.504	10.991	1.099
$\beta_n = 35^\circ$								
8	28.573	0.632	34.941	0.546	34.331	0.636	40.983	0.695
9	20.248	0.843	36.910	0.550	42.710	0.579	40.673	0.721
10	28.013	0.720	32.902	0.624	43.890	0.638	38.462	0.806
$\beta_n = 45^\circ$								
8	32.972	0.426	23.998	0.690	24.003	0.750	36.209	0.623
9	41.611	0.402	35.961	0.516	28.673	0.778	30.219	0.864
10	54.378	0.358	43.65	0.506	32.015	0.787	42.941	0.683
$\varepsilon_{\beta n} = 2.15...2.35$								
$\beta_n = 25^\circ$								
8	4.286	0.544	10.000	0	5.024	0.600	24.321	0.711
9	20.000	0	15.000	0	59.717	-0.600	9.029	1.181
10	58.639	-0.291	63.813	-0.407	44.878	-0.351	20.095	0.600
$\beta_n = 35^\circ$								
8	40.811	0.390	48.251	0.246	22.044	0.704	27.432	0.752
9	39.348	0.502	39.154	0.468	30.160	0.586	31.131	0.716
10	40.925	0.584	57.088	0.365	45.610	0.442	40.523	0.625
$\beta_n = 45^\circ$								
8	19.105	0.653	22.720	0.600	23.484	0.683	24.863	0.764
9	24.344	0.600	29.303	0.518	30.674	0.600	26.783	0.780
10	32.845	0.525	35.416	0.494	38.379	0.534	33.567	0.708

APPENDIX 1

ALGORITHM OF CALCULATING THE PRECISE VALUES OF REDUCED CURVATURE RADII AND THE MAJOR AXIS ANGLES OF INCLINATION OF THE INSTANT ELLIPTICAL CONTACT AREA TO TOOTH LINES OF CYLINDRICAL NOVIKOV GEARS

The precise values ρ_α^*, ρ_β^* and θ should be calculated separately for the point of contact of the gear's 1 addendum with the gear's 2 dedendum and for the contact point of the gear's 2 addendum with the gear's 1 dedendum.

In the first case the output parameters are marked with the superscript *(1)*, in the second – with the superscript *(2)*.

The algorithm of calculation for the first pair of contacting surfaces consists of the following operations:

$$c_1 = -(x_a^* - x_1^*)\cos\beta/(\rho_a^* \sin\alpha_k);$$

$$c_2 = -z_1 \sin\alpha_k/(2\cos\beta);$$

$$c_3 = \rho_a^* - (x_a^* - x_1^*)/\sin\alpha_k;$$

$$c_4 = c_2 + c_3(c_1\cos\beta - \sin^2\alpha_k \sin^2\beta);$$

$$c_5 = [c_4 - \rho_a^*(c_1^2 + \sin^2\alpha_k \sin^2\beta)]/(2\rho_a^* c_4);$$

$$c_6 = -\sin^2\alpha_k \sin^2\beta/(\rho_a^* c_4);$$

$$c_7 = (c_5^2 - c_6)^{0.5};$$

$$c_8 = (c_5 + c_7)^{-1};$$

$$c_9 = (c_5 - c_7)^{-1};$$

$$c_{10} = (\overset{*}{x}_f + \overset{*}{x}_2)\cos\beta / (\overset{*}{\rho}_f \sin\alpha_k);$$

$$c_{11} = z_2 \sin\alpha_k / (2\cos\beta);$$

$$c_{12} = \overset{*}{\rho}_f - (\overset{*}{x}_f + \overset{*}{x}_2)/\sin\alpha_k;$$

$$c_{13} = c_{11} - c_{12}(c_{10}\cos\beta + \sin^2\alpha_k \sin^2\beta);$$

$$c_{14} = [c_{13} - \overset{*}{\rho}_f(c_{10}^2 + \sin^2\alpha_k \sin^2\beta)]/(2\overset{*}{\rho}_f c_{13});$$

$$c_{15} = -\sin^2\alpha_k \sin^2\beta / (\overset{*}{\rho}_f c_{13});$$

$$c_{16} = (c_{14}^2 - c_{15})^{0.5};$$

$$c_{17} = (c_{14} + c_{16})^{-1};$$

$$c_{18} = (c_{14} - c_{16})^{-1};$$

$$c_{19} = -\overset{*}{\rho}_a(\sin^2\alpha_k \sin^2\beta - c_1^2) - c_4;$$

$$c_{20} = -\overset{*}{\rho}_f(\sin^2\alpha_k \sin^2\beta - c_{10}^2) - c_{13};$$

$$c_{21} = \arctan(2\overset{*}{\rho}_a c_1 \sin\alpha_k \sin\beta / c_{19});$$

$$c_{22} = \arctan(-2\overset{*}{\rho}_f c_{10} \sin\alpha_k \sin\beta / c_{20});$$

$$c_{23} = \cos(c_{21} - c_{22});$$

$$c_{24} = 2(c_7^2 + c_{16}^2 - 2c_7 c_{16} c_{23})^{0.5};$$

$$c_{25} = c_{22} - c_{21};$$

$$c_{26} = 0.5\arctan[c_{16}\sin c_{25} / (c_{16}\cos c_{25} - c_7)];$$

$$c_{27} = \overset{*}{\rho}_a \sin\alpha_k - \overset{*}{x}_a + \overset{*}{x}_1;$$

$$c_{28} = \overset{*}{\rho}_f \sin\alpha_k - \overset{*}{x}_f - \overset{*}{x}_2;$$

$$c_{29} = \arctan\{c_{27}\sin\beta\cos\beta / [\sin\alpha_k(0.5z_1\cos\beta + c_{27}\sin^2\beta)]\};$$

$$c_{30} = \arctan\{c_{28}\sin\beta\cos\beta / [\sin\alpha_k(-0.5z_2\cos\beta + c_{28}\sin^2\beta)]\}.$$

Now we determine the reduced main curvature radii: $\rho_\alpha^{*(1)} = \left|(c_5 - c_{14} + 0.5c_{24})^{-1}\right|$; $\rho_\beta^{*(2)} = \left|(c_5 - c_{14} - 0.5c_{24})^{-1}\right|$, and also the angles of inclination of the contact area to lines of the tooth correspondingly of the gear's 1 addendum and the gear's 2 dedendum: $\theta_1^{(1)} = 0.5c_{21} + c_{26} - c_{29}$; $\theta_2^{(1)} = 0.5c_{21} + c_{26} - c_{30}$.

Substituting in the given algorithm $\overset{*}{x}_1$ for $\overset{*}{x}_2$, z_1 for z_2 and, otherwise, $\overset{*}{x}_2$ for $\overset{*}{x}_1$ and z_2 for z_1, we will get the corresponding values of $\rho_\alpha^{*(2)}$, $\rho_\beta^{*(2)}$, $\theta_1^{(2)}$ and $\theta_2^{(2)}$ for the second pair of contacting surfaces.

APPENDIX 2

TABLES FOR DEFINITION OF COEFFICIENTS, CHARACTERIZING THE REDUCTION OF THE TEETH ACTIVE HEIGHT AS THE RESULT OF THEIR GENERATION

Table 1. Coefficients K_A, K_P, K_F, K_I for gears, made on the basis of the basic rack profile RGU-5.

$x^* \rightarrow$ / $z_v \downarrow$		-1.2	-1.1	-1.0	-0.9	-0.8	-0.7	-0.6	-0.5	-0.4	-0.3	-0.2	-0.1	0
9	K_A	-	-	-	-	-	-	-	471	468	458	439	412	378
	K_P	-	-	-	-	-	-	-	372	473	564	625	656	654
	K_F	-	-	-	-	-	-	-	298	322	307	359	406	429
	K_I	-	-	-	-	-	-	-	581	549	523	501	485	474
10	K_A	-	-	-	-	-	-	-	469	467	457	440	416	385
	K_P	-	-	-	-	-	-	-	382	481	563	619	646	645
	K_F	-	-	-	-	-	-	-	327	349	339	384	424	441
	K_I	-	-	-	-	-	-	-	577	549	525	506	491	481
12	K_A	-	-	-	-	-	-	-	466	465	456	442	422	396
	K_P	-	-	-	-	-	-	-	406	493	562	608	631	630
	K_F	-	-	-	-	-	-	-	371	389	385	430	449	460
	K_I	-	-	-	-	-	-	-	572	548	529	513	500	492
14	K_A	-	-	-	-	-	-	-	465	463	456	444	427	404
	K_P	-	-	-	-	-	-	-	424	502	562	601	621	620
	K_F	-	-	-	-	-	-	-	402	383	418	453	466	473
	K_I	-	-	-	-	-	-	-	568	548	531	517	507	500
17	K_A	-	-	-	-	-	-	460	463	461	456	445	431	413
	K_P	-	-	-	-	-	-	373	446	512	561	593	609	609
	K_F	-	-	-	-	-	-	424	435	425	450	475	483	487
	K_I	-	-	-	-	-	-	584	564	548	534	522	514	508
20	K_A	-	-	-	-	-	-	459	461	460	455	447	435	419
	K_P	-	-	-	-	-	-	395	463	518	560	588	602	601
	K_F	-	-	-	-	-	-	449	427	453	482	490	495	496
	K_I	-	-	-	-	-	-	578	562	548	536	526	519	514
25	K_A	-	-	-	-	-	453	458	460	459	455	448	438	426
	K_P	-	-	-	-	-	362	426	481	526	560	582	593	592
	K_F	-	-	-	-	-	470	443	466	483	502	506	508	507
	K_I	-	-	-	-	-	587	572	559	547	538	530	524	520
30	K_A	-	-	-	-	-	453	457	459	458	455	449	441	430
	K_P	-	-	-	-	-	392	447	494	531	559	578	587	586
	K_F	-	-	-	-	-	453	473	489	510	515	516	515	514
	K_I	-	-	-	-	-	580	568	557	547	539	533	528	525

Table 1. (Continued)

x^* →		-1.2	-1.1	-1.0	-0.9	-0.8	-0.7	-0.6	-0.5	-0.4	-0.3	-0.2	-0.1	0
z_v ↓														
40	K_A	-	-	-	-	448	453	456	458	457	455	450	444	436
	K_P	-	-	-	-	384	432	475	510	538	559	572	579	579
	K_F	-	-	-	-	480	496	509	527	528	528	527	523	522
	K_I	-	-	-	-	582	572	562	554	547	541	536	533	530
50	K_A	-	-	-	444	449	453	456	457	456	454	451	446	440
	K_P	-	-	-	375	418	457	491	519	542	558	569	575	575
	K_F	-	-	-	495	509	518	535	537	537	536	532	528	528
	K_I	-	-	-	584	575	567	559	552	547	542	538	535	533
60	K_A	-	-	440	445	450	454	456	456	456	454	451	447	442
	K_P	-	-	362	403	441	474	502	525	544	558	567	572	572
	K_F	-	-	505	516	525	542	544	544	542	540	535	532	531
	K_I	-	-	587	578	570	563	557	551	547	543	540	537	535
70	K_A	-	-	442	447	450	453	455	456	455	454	452	448	444
	K_P	-	-	389	425	457	486	510	530	546	558	566	570	569
	K_F	-	-	522	532	541	549	552	548	547	541	537	534	534
	K_I	-	-	581	573	567	561	555	551	547	543	541	538	537
80	K_A	-	-	443	447	451	453	455	456	455	454	452	449	445
	K_P	-	-	410	442	470	495	516	533	547	558	565	568	568
	K_F	-	-	533	539	553	553	552	551	550	543	539	536	536
	K_I	-	-	576	570	564	559	554	550	547	544	541	539	538
90	K_A	-	440	444	448	451	453	455	456	455	454	452	450	446
	K_P	-	395	426	454	480	501	520	536	548	558	564	567	567
	K_F	-	537	541	556	557	556	554	553	549	544	540	538	537
	K_I	-	579	573	567	562	558	553	550	547	544	542	540	539
100	K_A	436	441	445	449	451	453	455	455	455	454	452	450	447
	K_P	380	411	439	465	487	507	524	538	549	558	563	566	566
	K_F	538	544	553	559	561	558	557	556	549	544	541	539	538
	K_I	582	576	570	565	561	556	553	549	547	544	542	541	540

x^* →		0.1	0.2	0.3	0.4	0.5	0.6	0.7	0.8	0.9	1.0	1.0	1.2
z_v ↓													
9	K_A	336	286	229	166	095	-	-	-	-	-	-	-
	K_P	622	557	460	332	171	-	-	-	-	-	-	-
	K_F	444	476	525	591	655	-	-	-	-	-	-	-
	K_I	468	466	470	479	494	-	-	-	-	-	-	-
10	K_A	347	302	251	193	129	-	-	-	-	-	-	-
	K_P	615	557	470	354	221	-	-	-	-	-	-	-
	K_F	455	483	527	587	648	-	-	-	-	-	-	-
	K_I	475	474	478	486	499	-	-	-	-	-	-	-
12	K_A	364	327	283	234	180	-	-	-	-	-	-	-
	K_P	605	557	484	388	271	-	-	-	-	-	-	-
	K_F	471	494	531	581	636	-	-	-	-	-	-	-
	K_I	487	486	489	496	507	-	-	-	-	-	-	-
14	K_A	377	344	307	265	218	-	-	-	-	-	-	-
	K_P	599	557	495	412	309	-	-	-	-	-	-	-
	K_F	482	502	534	577	626	-	-	-	-	-	-	-
	K_I	496	495	498	503	513	-	-	-	-	-	-	-

$x^* \to$		0.1	0.2	0.3	0.4	0.5	0.6	0.7	0.8	0.9	1.0	1.0	1.2
$z_v \downarrow$													
	K_A	390	363	332	297	258	215	-	-	-	-	-	-
17	K_P	591	557	506	438	353	261	-	-	-	-	-	-
	K_F	494	510	536	572	617	652	-	-	-	-	-	-
	K_I	505	504	506	511	519	529	-	-	-	-	-	-
	K_A	399	376	350	320	286	249	-	-	-	-	-	-
20	K_P	586	557	513	456	384	298	-	-	-	-	-	-
	K_F	502	516	538	568	606	640	-	-	-	-	-	-
	K_I	511	510	512	516	523	531	-	-	-	-	-	-
	K_A	410	391	370	346	319	289	257	-	-	-	-	-
25	K_P	580	557	522	476	418	349	270	-	-	-	-	-
	K_F	511	523	540	565	595	626	652	-	-	-	-	-
	K_I	518	518	519	522	527	534	543	-	-	-	-	-
	K_A	417	402	384	364	341	316	289	-	-	-	-	-
30	K_P	576	557	528	489	441	384	317	-	-	-	-	-
	K_F	518	527	542	562	587	618	640	-	-	-	-	-
	K_I	523	522	524	526	531	536	544	-	-	-	-	-
	KA	426	414	401	386	369	350	329	307	-	-	-	-
40	KP	571	557	535	506	470	427	377	320	-	-	-	-
	KF	525	533	544	559	578	600	625	640	-	-	-	-
	KI	529	528	529	531	535	539	544	551	-	-	-	-
	KA	432	422	412	399	386	370	354	336	317	-	-	-
50	KP	569	557	540	516	488	453	413	367	316	-	-	-
	KF	530	536	545	557	572	590	611	627	642	-	-	-
	KI	532	532	533	534	537	540	545	550	556	-	-	-
	KA	435	428	419	408	397	384	371	355	339	322	-	-
60	KP	567	557	542	523	499	470	437	399	356	308	-	-
	KF	533	538	545	555	568	584	602	618	632	644	-	-
	KI	535	534	535	536	538	541	545	549	555	561	-	-
	KA	438	431	424	415	405	394	382	369	355	340	-	-
70	KP	565	557	544	528	507	483	454	421	384	343	-	-
	KF	536	540	546	554	565	579	594	612	624	644	-	-
	KI	536	536	537	538	540	542	545	549	553	559	-	-
	KA	440	434	427	420	441	402	391	380	367	354	-	-
80	KP	564	557	546	532	514	492	467	438	406	370	-	-
	KF	537	541	546	554	563	575	588	604	621	628	-	-
	KI	537	537	538	539	540	543	545	549	553	557	-	-
	KA	442	436	430	423	416	407	398	388	377	365	353	-
90	KP	563	557	547	534	518	499	477	451	423	391	356	-
	KF	538	542	547	553	562	572	584	598	613	630	631	-
	KI	538	538	539	540	541	543	545	548	552	556	560	-
	KA	443	438	433	426	420	412	403	394	385	374	363	351
100	KP	563	557	548	537	522	505	485	462	436	408	376	342
	KF	540	542	547	553	560	570	581	593	607	622	639	634
	KI	539	539	540	540	542	543	546	548	551	555	559	563

Note : 1. For gears with z_v > 100 all coefficients should be taken the same as for the gear with z_v = 100. 2. The values of coefficients for intermediate parameters z_v, x^* should be determined by means of linear interpolation.

Table 2. Coefficients K_A, K_P, K_F, K_I for gears, made on the basis of the basic rack profile according to the standard GOST 15023-76

$x^* \rightarrow$		-0.45	-0.4	-0.3	-0.2	-0.1	0
$z_v \downarrow$							
9	K_A	-	-	578	554	518	469
	K_P	-	-	309	499	542	450
	K_F	-	-	567	497	480	413
	K_I	-	-	725	669	623	588
10	K_A	-	-	577	556	523	478
	K_P	-	-	317	495	533	450
	K_F	-	-	598	524	494	424
	K_I	-	-	713	663	622	590
12	K_A	-	-	576	557	530	493
	K_P	-	-	333	487	519	450
	K_F	-	-	641	574	514	439
	K_I	-	-	696	654	620	593
14	K_A	-	-	575	559	535	503
	K_P	-	-	345	482	509	450
	K_F	-	-	670	607	527	449
	K_I	-	-	683	648	618	595
17	K_A	-	-	574	561	541	514
	K_P	-	-	364	476	499	450
	K_F	-	-	695	622	534	459
	K_I	-	-	670	641	617	598
20	K_A	-	-	573	562	545	522
	K_P	-	-	394	472	492	450
	K_F	-	-	692	626	537	465
	K_I	-	-	661	636	615	599
25	K_A	-	-	572	563	550	532
	K_P	-	-	408	468	483	450
	K_F	-	-	712	624	532	473
	K_I	-	-	651	631	614	601
30	K_A	-	-	571	564	553	538
	K_P	-	-	414	465	478	450
	K_F	-	-	710	616	527	479
	K_I	-	-	644	627	613	603
40	K_A	-	573	570	565	557	545
	K_P	-	362	421	461	471	450
	K_F	-	787	685	588	521	485
	K_I	-	650	635	623	612	604
50	K_A	-	572	570	566	559	550
	K_P	-	376	427	459	467	450
	K_F	-	750	649	572	518	489
	K_I	-	642	630	620	612	605
60	K_A	-	572	570	566	560	553
	K_P	-	385	431	457	464	450
	K_F	-	710	625	560	516	492
	K_I	-	636	626	618	611	606
70	K_A	-	571	570	567	562	555
	K_P	-	404	434	456	462	450
	K_F	-	681	608	552	514	493
	K_I	-	632	624	617	611	606

$x^* \rightarrow$		-0.45	-0.4	-0.3	-0.2	-0.1	0
$z_v \downarrow$							
80	K_A	-	571	569	567	562	557
	K_P	-	400	436	456	460	450
	K_F	-	659	595	546	513	495
	K_I	-	630	622	616	611	607
90	K_A	571	571	569	567	563	558
	K_P	395	406	437	455	459	450
	K_F	675	642	585	542	512	496
	K_I	631	627	621	615	611	607
100	K_A	570	570	569	567	564	559
	K_P	392	410	438	454	458	450
	K_F	658	628	577	538	511	497
	K_I	629	625	620	615	610	607

$x^* \rightarrow$		0.02	0.04	0.06	0.08	0.1	0.15	0.2	0.25	0.3
$z_v \downarrow$										
9	K_A	457	445	433	-	-	-	-	-	-
	K_P	415	375	334	-	-	-	-	-	-
	K_F	401	390	385	-	-	-	-	-	-
	K_I	582	576	571	-	-	-	-	-	-
10	K_A	468	457	446	435	-	-	-	-	-
	K_P	419	382	341	302	-	-	-	-	-
	K_F	411	402	397	397	-	-	-	-	-
	K_I	584	579	575	570	-	-	-	-	-
12	K_A	484	475	466	456	-	-	-	-	-
	K_P	424	394	359	329	-	-	-	-	-
	K_F	427	419	415	415	-	-	-	-	-
	K_I	588	584	580	577	-	-	-	-	-
14	K_A	496	488	480	472	-	-	-	-	-
	K_P	428	402	372	339	-	-	-	-	-
	K_F	438	431	428	428	-	-	-	-	-
	K_I	591	588	585	581	-	-	-	-	-
17	K_A	509	502	496	489	481	-	-	-	-
	K_P	431	410	386	359	332	-	-	-	-
	K_F	450	444	441	441	444	-	-	-	-
	K_I	595	592	589	586	584	-	-	-	-
20	K_A	517	512	506	501	494	-	-	-	-
	K_P	434	416	396	373	347	-	-	-	-
	K_F	458	453	451	451	453	-	-	-	-
	K_I	597	594	592	590	588	-	-	-	-
25	K_A	527	523	519	514	509	496	-	-	-
	K_P	437	423	406	388	368	317	-	-	-
	K_F	467	464	462	462	464	477	-	-	-
	K_I	599	597	595	594	592	589	-	-	-
30	K_A	534	530	527	523	519	508	-	-	-
	K_P	439	427	414	398	381	340	-	-	-
	K_F	474	470	469	469	471	482	-	-	-
	K_I	601	599	598	596	595	592	-	-	-

Table 2. (Continued)

$x^* \rightarrow$ / $z_v \downarrow$		0.02	0.04	0.06	0.08	0.1	0.15	0.2	0.25	0.3
40	K_A	543	540	537	534	531	523	514	-	-
	K_P	442	433	423	411	399	361	323	-	-
	K_F	481	479	478	478	479	488	504	-	-
	K_I	603	602	601	599	598	596	595	-	-
50	K_A	548	546	543	541	539	532	525	517	-
	K_P	444	436	428	419	409	379	352	308	-
	K_F	486	484	483	483	484	491	504	523	-
	K_I	604	603	602	601	601	599	598	597	-
60	K_A	551	549	548	545	544	538	532	526	-
	K_P	445	439	432	424	416	391	361	328	-
	K_F	489	488	487	487	488	493	504	520	-
	K_I	605	604	603	603	602	601	600	599	-
70	K_A	554	552	550	549	547	542	537	532	-
	K_P	445	440	434	428	421	399	374	350	-
	K_F	491	490	489	489	490	495	504	518	-
	K_I	606	605	604	604	603	602	601	601	-
80	K_A	555	554	553	551	550	546	541	536	-
	K_P	446	441	436	431	424	406	383	357	-
	K_F	493	492	491	491	492	496	504	516	-
	K_I	606	605	605	604	604	603	602	602	-
90	K_A	557	556	554	553	552	548	544	540	535
	K_P	446	442	438	433	427	411	391	367	341
	K_F	494	493	493	493	493	497	504	515	529
	K_I	606	606	605	605	604	603	603	602	602
100	K_A	558	557	556	554	553	550	546	543	539
	K_P	447	443	439	434	429	415	397	376	352
	K_F	495	494	494	494	494	498	504	514	526
	K_I	607	606	606	605	605	604	603	603	603

Note :

1. For gears with $z_v > 100$ all coefficients should be taken the same as for the gear with $z_v = 100$.
2. The values of coefficients for intermediate parameters z_v, x^* should be determined by means of linear interpolation.

Table 3. Coefficients for gears, made on the basis of the basic rack profile according to the standard GOST 30224-96

$x^* \rightarrow$ / $z_v \downarrow$		-0.9	-0.8	-0.7	-0.6	-0.5	-0.4	-0.3	-0.2	-0.1	0
9	K_A	-	-	-	-	-	517	508	489	464	430
	K_P	-	-	-	-	-	439	512	570	588	563
	K_F	-	-	-	-	-	481	474	477	494	520
	K_I	-	-	-	-	-	576	551	530	515	504
10	K_A	-	-	-	-	-	516	507	491	467	436
	K_P	-	-	-	-	-	443	510	564	580	557
	K_F	-	-	-	-	-	496	493	498	512	536
	K_I	-	-	-	-	-	577	554	535	521	512
12	K_A	-	-	-	-	-	513	506	492	472	447
	K_P	-	-	-	-	-	448	509	555	568	549
	K_F	-	-	-	-	-	526	521	527	538	559
	K_I	-	-	-	-	-	578	559	543	531	523
14	K_A	-	-	-	-	-	511	505	493	476	454
	K_P	-	-	-	-	-	453	509	548	559	543
	K_F	-	-	-	-	-	546	545	548	557	574
	K_I	-	-	-	-	-	578	562	549	539	532
17	K_A	-	-	-	-	510	509	504	494	480	462
	K_P	-	-	-	-	408	460	508	541	550	537
	K_F	-	-	-	-	575	568	566	568	574	590
	K_I	-	-	-	-	595	579	565	554	546	541
20	K_A	-	-	-	-	509	508	503	495	483	468
	K_P	-	-	-	-	417	465	508	536	544	532
	K_F	-	-	-	-	590	584	581	584	590	602
	K_I	-	-	-	-	593	579	568	558	551	547
25	K_A	-	-	-	-	507	506	503	496	486	474
	K_P	-	-	-	-	430	471	508	530	536	527
	K_F	-	-	-	-	606	603	599	602	608	614
	K_I	-	-	-	-	591	580	571	563	558	554
30	K_A	-	-	-	504	506	505	502	497	489	478
	K_P	-	-	-	400	439	477	508	526	532	524
	K_F	-	-	-	622	618	613	610	613	620	623
	K_I	-	-	-	600	589	580	572	566	562	558
40	K_A	-	-	-	503	505	504	502	498	492	484
	K_P	-	-	-	420	453	484	508	521	525	520
	K_F	-	-	-	635	630	627	626	626	632	633
	K_I	-	-	-	595	587	580	575	570	567	564
50	K_A	-	-	500	502	504	503	501	498	493	487
	K_P	-	-	404	433	462	489	508	519	522	517
	K_F	-	-	649	641	638	636	636	636	639	640
	K_I	-	-	600	593	586	581	576	572	570	568
60	K_A	-	-	500	502	503	503	501	498	494	489
	K_P	-	-	416	443	470	492	507	517	519	516
	K_F	-	-	652	647	644	642	638	644	643	644
	K_I	-	-	597	591	585	581	577	574	572	570
70	K_A	-	497	500	502	503	502	501	499	495	490
	K_P	-	401	427	451	475	494	507	515	518	514
	K_F	-	659	655	652	648	648	644	648	646	647
	K_I	-	601	595	589	585	581	578	575	573	572

Table 3. Continued

$x^* \rightarrow$		-0.9	-0.8	-0.7	-0.6	-0.5	-0.4	-0.3	-0.2	-0.1	0
$z_v \downarrow$											
80	K_A	-	497	500	502	502	502	501	499	496	492
	K_P	-	411	435	458	479	496	507	514	516	513
	K_F	-	662	658	654	651	650	648	650	649	649
	K_I	-	598	593	588	584	581	578	576	574	573
90	K_A	495	498	500	501	502	502	501	499	496	493
	K_P	398	419	441	463	482	497	507	513	515	513
	K_F	669	663	659	653	653	648	652	652	650	651
	K_I	601	596	592	588	584	581	579	576	575	574
100	K_A	495	498	500	501	502	502	501	499	497	493
	K_P	405	426	447	468	485	498	507	513	514	512
	K_F	669	666	660	657	655	651	654	653	652	652
	K_I	599	595	591	587	584	581	579	577	576	575

$x^* \rightarrow$		0.05	0.1	0.15	0.2	0.25	0.3	0.35	0.4	0.45	0.5	0.55	0.6	0.65
$z_v \downarrow$														
9	K_A	410	388	365	-	-	-	-	-	-	-	-	-	-
	K_P	534	494	445	-	-	-	-	-	-	-	-	-	-
	K_F	534	550	571	-	-	-	-	-	-	-	-	-	-
	K_I	501	499	498	-	-	-	-	-	-	-	-	-	-
10	K_A	419	399	378	-	-	-	-	-	-	-	-	-	-
	K_P	531	496	453	-	-	-	-	-	-	-	-	-	-
	K_F	547	561	581	-	-	-	-	-	-	-	-	-	-
	K_I	509	507	506	-	-	-	-	-	-	-	-	-	-
12	K_A	432	415	397	378	-	-	-	-	-	-	-	-	-
	K_P	527	498	461	415	-	-	-	-	-	-	-	-	-
	K_F	567	579	595	615	-	-	-	-	-	-	-	-	-
	K_I	521	519	519	519	-	-	-	-	-	-	-	-	-
14	K_A	441	427	412	395	-	-	-	-	-	-	-	-	-
	K_P	524	499	467	431	-	-	-	-	-	-	-	-	-
	K_F	581	591	605	622	-	-	-	-	-	-	-	-	-
	K_I	530	528	528	528	-	-	-	-	-	-	-	-	-
17	K_A	451	440	427	413	-	-	-	-	-	-	-	-	-
	K_P	521	500	474	443	-	-	-	-	-	-	-	-	-
	K_F	596	604	615	630	-	-	-	-	-	-	-	-	-
	K_I	539	538	537	537	-	-	-	-	-	-	-	-	-
20	K_A	458	449	438	426	413	-	-	-	-	-	-	-	-
	K_P	519	501	479	451	423	-	-	-	-	-	-	-	-
	K_F	606	603	623	635	650	-	-	-	-	-	-	-	-
	K_I	545	544	544	544	545	-	-	-	-	-	-	-	-
25	K_A	467	459	450	441	430	419	-	-	-	-	-	-	-
	K_P	517	503	484	462	437	410	-	-	-	-	-	-	-
	K_F	618	624	631	641	653	666	-	-	-	-	-	-	-
	K_I	553	552	551	552	552	553	-	-	-	-	-	-	-

$x^* \rightarrow$ / $z_v \downarrow$		0.05	0.1	0.15	0.2	0.25	0.3	0.35	0.4	0.45	0.5	0.55	0.6	0.65
30	K_A	472	466	458	450	442	433	-	-	-	-	-	-	-
	K_P	515	503	488	470	449	425	-	-	-	-	-	-	-
	K_F	626	630	637	645	655	666	-	-	-	-	-	-	-
	K_I	557	557	556	557	557	558	-	-	-	-	-	-	-
40	K_A	479	474	468	462	456	449	442	-	-	-	-	-	-
	K_P	513	504	493	479	463	445	424	-	-	-	-	-	-
	K_F	635	639	644	650	657	666	676	-	-	-	-	-	-
	K_I	563	563	563	563	563	564	565	-	-	-	-	-	-
50	K_A	483	479	475	470	465	459	454	447	-	-	-	-	-
	K_P	512	505	496	485	472	457	440	424	-	-	-	-	-
	K_F	641	644	648	653	659	666	673	682	-	-	-	-	-
	K_I	567	567	567	567	567	567	568	569	-	-	-	-	-
60	K_A	486	483	479	475	471	466	461	456	450	-	-	-	-
	K_P	511	505	498	488	478	465	451	436	418	-	-	-	-
	K_F	645	648	651	655	660	665	672	679	687	-	-	-	-
	K_I	569	569	569	569	569	570	570	571	572	-	-	-	-
70	K_A	488	485	482	479	475	471	467	462	457	452	-	-	-
	K_P	511	506	499	491	482	471	459	446	432	415	-	-	-
	K_F	648	650	653	656	660	665	671	677	684	690	-	-	-
	K_I	571	571	571	571	571	571	572	573	574	575	-	-	-
80	K_A	490	487	484	481	478	474	471	467	463	458	454	-	-
	K_P	510	506	500	493	485	476	465	453	441	426	413	-	-
	K_F	650	652	654	657	661	665	670	676	682	688	692	-	-
	K_I	573	572	572	572	572	573	573	574	575	575	577	-	-
90	K_A	491	488	486	484	481	477	474	471	467	463	459	455	-
	K_P	510	506	501	495	487	479	470	459	448	435	421	410	-
	K_F	652	653	655	658	661	665	670	675	680	686	690	693	-
	K_I	574	573	573	573	573	574	574	575	575	576	577	578	-
100	K_A	491	490	488	485	482	480	477	474	470	467	463	459	455
	K_P	510	506	501	496	489	482	474	464	454	443	431	418	407
	K_F	653	654	656	659	662	665	669	674	678	684	689	692	694
	K_I	574	574	574	574	574	575	575	575	576	577	578	578	580

Note :

1. For gears with $z_v > 100$ all coefficients should be taken the same as for the gear with $z_v = 100$.
2. The values of coefficients for intermediate parameters z_v, x^* should be determined by means of linear interpolation.

APPENDIX 3

EXAMPLES OF GEOMETRY AND STRENGTH ENGINEERING ANALYSIS OF CYLINDRICAL NOVIKOV GEARING ACCORDING TO TECHNIQUES, DESCRIBED IN THE MONOGRAPH

Example 1

It is necessary to perform the design and also checking geometrical and strength analysis with control of engagement quality of the helical Novikov gearing with the basic rack profile according to the standard GOST 15023-76 for the following initial data: operating load $T_2 = 30000\,N \cdot m$, peak instantaneous load $T_{max\,2} = 60000\,N \cdot m$, gear ratio of the pair $u = 5$, pinion - nitrided, with the surface hardness $H_{HB1} = HB\ 550$, the core hardness $H_k = HB\ 300$, gear – thermally improved, with the tooth hardness $H_{HB2} = HB\ 250$, $n_1 = 50\ min^{-1}$, calculated operating time of the pair – *25 thousand hours*, operating mode – constant, reversible, reversion characteristics (ratio between diagrams of torques at the opposite and main tooth flanks) $b_\theta = 0.5$, degrees of accuracy (standard GOST 1643-81): according to kinematics $k=9$, smoothness $k_p = 8$, contact $k_k = 8$, the accepted value of the "risk level" $t_c = 5\%$, the angle $\beta = 21^\circ$, the proposed axial overlap ratio $\varepsilon_\beta \approx 1.7$.

Design Analysis

Accepting $z_\Sigma = 84$ (Chapter 16) according to (16.1) we determine $z_1 = 84/(5+1) = 14$, $z_2 = u \cdot z_1 = 70$.

Then we define the following parameters:

1) According to (2.20) $\rho^*_\beta = 70[2(5+1)sin27^\circ sin^2 21^\circ cos21^\circ] = 107$.
2) According to (3.18) $l^* = 2 \cdot 1.15 sin[0.5(51.5^\circ - 8.6^\circ)] = 0.841$, where according to (3.19) $\alpha_a = \arcsin(0.9 / 1.15) = 51.5^\circ$.
3) According to (11.9) $Z_l = 0.95 - 0.00037(250 - 200) = 0.932$.

4) Calculating according to [43] and recommendations for S_H (Chapter 14) the minimum value for the pinion and gear $\sigma_{HP2} = 570\ MPa$, according to (16.2) we obtain:

$$m_H = (3960/570)^{0.485}\left[30000/(70\cos 27^\circ)\right]^{0.333} \cdot 107^{-0.151}(0.932 \cdot 0.841)^{-0.515} = 11.2mm.$$

5) According to the diagram from Fig. 38 we determine (for $x^*_{1,2} = 0$): for the pinion $(z_{v1} \approx 17)$ $Y_{v1} \approx 0.62$, for the gear $(z_{v2} \approx 86)$ $Y_{v2} \approx 0.59$; since, according to [43], $\sigma^0_{F\lim b1} = 290 + 1.2 \cdot 300 = 650\ MPa$, and $\sigma^0_{F\lim b2} = 1.75 \cdot 250 = 438\ MPa$, then it is evident, that the calculation of the bending endurance should also be performed for the gear as the weakest element.

6) Assigning $a^*_F \approx 2$, we define according to (13.3):

$$Y_a = 1 - 0.07 \cdot 2^2(1.19 - 3.42/86)/(1 + 0.14 \cdot 2 + 0.07 \cdot 2^2) = 0.794$$

7) Determining according to [43] and recommendations for S_F (Chapter 14) parameters: $Y_{A2} = 1 - 0.35 \cdot 0.5 = 0.825$, $S_F = 1.1$, we will obtain approximately $\sigma_{FP2} = 438 \cdot 0.825/1.1 = 329\ MPa$; then according to (16.3) we define:

$$m_F = 12.6\left[30000\,0.59 \cdot 0.794/(329 \cdot 70)\right]^{1/3} = 10.7mm$$

8) Now according to (16.4) we have the maximum calculated value $m = 11.2\ mm$; considering, that the assumed factor ε_β must be approximately $1.7 > 1.6$, according to the Note 1 in Chapter 16 we decrease the module by 10% and round it up to the standard value $m = 10\ mm$ [37].

Letus refine geometrical parameters.

According to (2.2) $a^* = (14+70)/(2\cos 21^\circ) = 44.988$, $a = ma^* = 10 \cdot 44.988 = 449.880mm$

Taking the standard value [41] $a_w = 450mm$ $(a^*_w = 450/10 = 45)$, on the basis of (2.9) we have: $x^*_\Sigma = \dfrac{\sqrt{3.851840(45^2 - 44.988^2)\cos^2 21^\circ + 45^2} - 44.988}{1 + 3.851840\cos^2 21^\circ} = 0.012.$

Assigning $x^*_1 = 0.082$ according to [213], we will obtain $x^*_2 = x^*_\Sigma - x^*_1 = 0.012 - 0.082 = -0.070$

On the basis of (16.5), the operating toothing width of the gear pair $b_w = \pi m \varepsilon_\beta / \sin 21^\circ = \pi \cdot 10 \cdot 1.7 / \sin 21^\circ = 149mm$

We round (it) up to $b_w = 150mm$. The refined axial overlap ratio $\varepsilon_\beta = 150\sin 21^\circ/(10\pi) = 1.711$.

The final main geometrical parameters of the designed gearing are: $a_w = 450mm$, $m = 10mm$, $z_1 = 14$, $z_2 = 70$, $\beta = 21^\circ$, $x^*_1 = 0.082$, $x^*_2 = -0.070$, $b_1 = 160mm$, $b_2 = 150mm$.

Geometrical Analysis

1) Pitch diameters (2.10): $d_1 = 10 \cdot 14/\cos 21^o = 149.960\ mm$, $d_2 = 10 \cdot 70/\cos 21^o = 749.801\ mm$

2) Reference diameters (2.11): d_{w1}=2·450/(5+1)= 150 mm, d_{w2}=2·450·5/(5+1)= 750 mm.
3) Diameters of tooth tips (2.12): d_{a1}=149.960+2·10(0.9+0.082)=169.600 mm, d_{a2}=749.801+2·10(0.9-0.070)=766.401 mm.
4) Root diameters (2.13): d_{f1}=149.960+2·10(0.082-1.05)=130.600 mm, d_{f2}=749.801+2·10(-0.070-1.05)=727.401 mm.
5) Auxiliary values (2.18) and the Table 5: A_q = 0.65232, B_q = 0.25875, C_q = $0.65232 \sin^2 21° + 0.25875 = 0.342$.
6) Phase axial overlap ratio (2.18): $\varepsilon_q = \max\{0.342,(1-0.342)\} = 0.658$.
7) Reduced tooth number of equivalent straight-tooth gears (2.22): z_{v1} $=14/\cos^3 21° = 17.2057$, $z_{v2} = 70/\cos^3 21° = 86.0286$.
8) Backlash in the gearing (2.26), (2.27):
$j_w = 10[0.050 + 4 \cdot \cot^2 27° \cdot 0.012^2/(17.2057 + 86.0286 + 2 \cdot 0.012)] = 0.500 mm$
9) Choosing $\vartheta = 27.5°$, we define (4.5): A_1^*=1.15 sin 27.5°+0.082+0.5·17.2057=9.215861, A_2^*=1.15 sin 27.5°-0.070+0.5·86.0286=43.475311, B_1^*=1.15 cos 27.5°+ 0.082/tan 27.5° =1.177583, B_2^*=1.15 cos 27.5° - 0.070/tan 27.5° =0.885594, φ_1=2(0.082/ tan 27.5° +0.3927)/17.2057=0.063958, φ_2=2(-0.070/ tan 27.5° +0.3927)/86.0286=0.006003, $\tilde{x}_1^*$ =9.215861·cos 0.063958+1.177583·sin 0.06395-0.5·17.2057=0.6694, $\tilde{x}_2^*$ =43.475311·cos 0.006003+0.885594·sin 0.006003-0.5·86.0286=0.4655, $\tilde{y}_1^*$ =1.177583·cos 0.063958-9.215861·sin 0.063958=0.5861, $\tilde{y}_2^*$ =0.885594·cos 0.006003-43.475311·sin 0.006003=0.6246.
10) Tooth thickness along the chord (4.7): S_{x1}=2·10·0.5861=11.722 mm, S_{x2} =2·10·0.6246=12.492 mm.
11) Height to the chord (4.6): h_{x1}=10(0.9+0.082-0.6694)=3.126 mm, h_{x2}=10(0.9-0.070-0.4655)=3.645mm.
12) Design angles (Chapter 4): $\alpha_{min} = 1.06 \cdot 8.384° = 8.887°$, $\alpha_{max} = 0.85 \cdot 51.5° = 43.775°$.
13) Minimum parametrical angle (4.8): φ_{min}= arctan(tan 8.887°/cos 21°)=0.165948 (rad.).
14) Maximum parametrical angle (4.8): φ_{max}= arctan (tan 43.775°/cos 21°)=0.798374 (rad.).
15) Minimum possible tooth number within the length of the common normal. Calculations according to (4.9) provide: $(z_{n1})_{min} = 3$, $(z_{n2})_{min} = 6$.
16) Maximum possible tooth number within the length of the common normal. Calculations according to (4.10) provide: $(z_{n1})_{max} = 5$, $(z_{n2})_{max} = 20$.

17) Current parametrical angle (4.11). Choosing for the pinion $z_{n1} = 4$, for the gear $z_{n2} = 10$, we obtain: $\varphi_1 = \frac{\pi(4-1)-0.7854}{14} - \frac{2\cdot 0.082}{14\cos 21^\circ \tan\varphi_1} - \frac{\sin(2\varphi_1)\tan^2 21^\circ}{2}$,

$\varphi_2 = \frac{\pi(10-1)-0.7854}{70} - \frac{2(-0.070)}{70\cos 21^\circ \tan\varphi_2} - \frac{\sin(2\varphi_2)\tan^2 21^\circ}{2}$. Solving the equations by iteration method according to recommendations in Chapter 4, we obtain: $\varphi_1 = 0.531380$ (rad.), $\varphi_2 = 0.350977$ (rad.).

18) Length of the common normal (4.12):

$$W_1 = 2\cdot 10\left[\left(\frac{14\sin 0.531380}{2\cos 21^0} + \frac{0.082}{\sin 0.531380}\right)\sqrt{1+\cos^2 0.531380\cdot\tan^2 21^0} + 1.15\right] = 106.450\,mm,$$

$$W_2 = 2\cdot 10\left[\left(\frac{70\sin 0.350977}{2\cos 21^0} - \frac{0.070}{\sin 0.350977}\right)\sqrt{1+\cos^2 0.350977\cdot\tan^2 21^0} + 1.15\right] = 292.701\,mm.$$

Conditions (4.13) $W_1 = 106450\,mm < b_1/\sin\beta = 44647\,mm$ and $W_2 = 292701\,mm < b_2/\sin\beta = 41856\,mm$ are satisfied, therefore, the common normal lines with the length W_1 and W_2 can be measured. Since teeth, made on the basis of the basic rack profile according to the standard GOST 15023-76, do not have the dedendum "recess", the condition (4.14) should not be checked.

Evaluation of Factors, Determining the Quality of Engagement

1) Minimum number of contact points, simultaneously participating in engagement. According to the Table 6 for $\varepsilon_\beta = 1.711$ and $\varepsilon_q = 0.658$ ($\varepsilon_q \neq 0.5$) we obtain $n_{min} = 3$.
2) Determining $L_1 = \sqrt{1+2\sin 8.384^\circ\cdot 17.2057/1.15} = 2.316$, $L_2 = \sqrt{1+2\sin 8.384^\circ\cdot 86.0286/1.15} = 4.776$, we obtain the minimum allowable calculated value of shift coefficients according to the condition of the absence of undercut (3.5) $x^*_{min\,1} = -0.5\cdot 1.15\sin 8.384^\circ(2.316+1) = -0.278$, $x^*_{min\,2} = -0.5\cdot 1.15\sin 8.384^\circ(4.776+1) = -0.484$ and the maximum allowable calculated value of shift coefficients $x^*_{max\,1} = 0.5\cdot 1.15\sin 8.384^\circ(2.316-1) = 0.110$, $x^*_{max\,2} = 0.5\cdot 1.15\sin 8.384^\circ(4.776-1) = 0.317$.
3) Check-up of the condition of the absence of undercut (3.6): $x^*_{min\,1} = -0.278 < x^*_1 = 0.082 < x^*_{max\,1} = 0.110$ and $x^*_{min\,2} = -0.484 < x^*_2 = -0.070 < x^*_{max\,2} = 0.317$. Shift coefficients x^*_1 and x^*_2 are located inside the areas of the absence of undercut. According to recommendations given in Chapter 3, it is not necessary to check the teeth made on the basis of the basic rack profile according to the standard GOST 15023-76 for the absence of tooth thinning by the condition (3.12). For the same reason, there is no point in checking the allowable location of the pitch point line in the gearing according to the condition (3.16), because the near-pitch-point conversion area is not eliminated from operation (see paragraph 3.3 of Chapter 3).

4) Auxiliary coefficients. According to the Table 2 of the Appendix 2 for z_{v1}=17.2057, x_1^*=0.082 and z_{v2}=86.0286, x_2^*=-0.070 we obtain by means of linear interpolation:

K_{A1}=489, K_{P1}=358, K_{F1}=442, K_{I1}=586,

K_{A2}=561, K_{P2}=457, K_{F2}=507, K_{I2}=610,

and according to the Table 17 $K_{KA}=1.015$, $K_{KP}=0.98$, $K_{KF}=0.95$, $K_{KI}=1.052$ Then (11.1), (11.2): $K_l^{(1)}=min\{1.015\cdot 489,\ 1.052\cdot 610\}+min\{0.98\cdot 358,\ 0.95\cdot 507\}=847,$

$K_l^{(2)}=min\{1.015\cdot 561,\ 1.052\cdot 586\}+min\{0.98\cdot 457,\ 0.95\cdot 442\}=989.$

5) Coefficient of the tooth active height reduction (4.10): $K_l=\{847,989\}\cdot 10^{-3}=0.847.$

Checking Strength Analysis

Letus consider two cases: the case, when the rigidity of shafts and supports is not considered (n_k = 0) and the case, when the rigidity of shafts and supports is considered (n_k =1) – see paragraph 10.2. of Chapter 10.

Bending Endurance

$$\sigma_{FPi}=\sigma^0_{F\,lim\,bi}Y_\delta Y_{xi}Y_{Ni}Y_A/S_{Fi}.$$

1) Allowable stress [43]:

$$\sigma^0_{F\,lim\,b1}=290+1.2H_k=290+1.2\cdot 300=650\ \ MPa,$$
$$\sigma^0_{F\,lim\,b2}=1.75\cdot H_{HB2}=1.75\cdot 250=438\ \ MPa;$$
$$Y_\delta=1.082-0.172\log m=1.082-0.172\log 10=0.91;$$
$$Y_{X1}=1.05-0.000125\,d_{w1}=1.05-0.000125\cdot 150=1.03,$$
$$Y_{X2}=1.05-0.000125\,d_{w2}=1.05-0.000125\cdot 750=0.96;$$

$$N_{F1}-60\cdot 50\cdot 25000=75\cdot 10^6\ \ (cycles\ ),$$
$$N_{F2}=N_{F1}/u=75\cdot 10^6/5=15\cdot 10^6\ \ (cycles\ );$$
$$Y_{N1}=\sqrt[9]{\frac{4\cdot 10^6}{75\cdot 10^6}}=0.722;\ \ we\ accept\ \ Y_{N1}=1;$$
$$Y_{N2}=\sqrt[6]{\frac{4\cdot 10^6}{15\cdot 10^6}}=0.802;\ \ we\ accept\ \ Y_{N2}=1;$$

according to conclusions of Chapter 14: $S_{F1}=1.1,\ \ S_{F2}=1.1;$ coefficient Y_A, considering the two-sided load application [43]:

$$Y_A=1-\gamma_A T'_{F2}/T_{F2}=1-\gamma_A b_\theta,$$
$$Y_{A1}=1-0.1\cdot 0.5=0.95,$$
$$Y_{A2}=1-0.35\cdot 0.5=0.825;$$

Now we have:

$$\sigma_{FP1}=650\cdot 0.91\cdot 1.03\cdot 1\cdot 0.95/1.1=526\ \ MPa,$$
$$\sigma_{FP2}=438\cdot 0.91\cdot 0.96\cdot 1\cdot 0.825/1.1=287\ \ MPa.$$

2) Acting stress. Comparing σ_{FP1} and σ_{FP2}, it is obvious, that the calculation should be performed only for the driven gear (the weakest element in our example). According to (13.1) $\sigma_{F2}=2000K_F T_2 Y_{v2}Y_{a2}/(m^3 z_2)$. Using the

diagram (Fig. 38) for $z_{v2} = z_2 / cos^3 \beta = 70 / cos^3 21° = 86$ and $x_2 = -0.070$ we determine $Y_{v2} = 0.59$.

According to (2.19) and (2.20) we define
$$\rho^*_\alpha = 1.15 \cdot 1.27 / 0.12 = 12.17,$$
$$\rho^*_\beta = 70 / [2(5+1) \sin 27° \sin^2 21° \cos 21°] = 107.17;$$
$$C_{\alpha\beta} = \rho^*_\alpha / \rho^*_\beta = 12.17 / 107.17 = 0.114;$$
by (11.27): $n_a = 0.97016 \cdot 0.114^{-0.39010} = 2.263$.

Determining the normal force (10.2) $F_n = 2000T_2 / (mz_2 \cos\alpha_k) = 2000 \cdot 30000 / (10 \cdot 70 \cdot \cos 27°) = 96199\,N$, according to (11.24), (11.25) we define $a_H = 2.263 [1.5 \cdot 8.47 \cdot 10^{-6} \cdot 96199 \cdot 12.17 \cdot 10 / (1+0.114)]^{1/3} = 11.6\ mm$.

Then $a^*_F = a^*_H = a_H / m = 1.16$; on the basis of (13.3) and Table 24:
$$Y_{a2} = 1 - \frac{0.07 \cdot 1.16^2}{1 + 0.14 \cdot 1.16 + 0.07 \cdot 1.16^2}(1.19 - 3.42 / 86) = 0.914.$$

The load coefficient K_F according to (10.8) is equal to $K_F = K_A K_{F\sigma} K_{Fv}$.

We accept [43] $K_A = 1$.

Then we define: according to (10.11) for $K_b = 1$: $W_\delta = 0.0692 \cdot 10^{-0.926} \, 30000^{0.642} (107.17)^{-0.299} \cdot 70^{-0.67} (\cos 27°)^{-1.642} (\cos 21°)^{-1} = 0.1141\,mm$, for the reference design value $\varepsilon_\beta = 1 + \varepsilon_q = 1.658$ we obtain $b_w = 145\ mm$ (see Note 1 to the Table 13), and then on the basis of [40] for $k_p = k_k = 8$ we have $F_{\beta 1} = 0.034\ mm$, $f_{pt1} = 0.032\ mm$; for $t_c = 5\%$ we get $\gamma_c = 1.96$ (Table 9) and then (10.15) $\lambda = 1.96\sqrt{0.034^2 + 0.032^2} / 0.1141 = 0.802$.

On the basis of the Table 13, for $n_k = 0$: $A = 0.847$, $a = 0.609$ and according to (10.14) $K^0_{F\sigma} = 0.847 \cdot 0.802^{0.609} = 0.741$.

Determining the tangential velocity $v = \frac{\pi d_{w1} n_1}{60000} = \frac{\pi 150 \cdot 50}{60000} = 0.39$ m/s and the run-in coefficient [43]: $K_w = 1 - \frac{20}{(0.01 \cdot HB_{min} + 2)^2 (v+4)^{0.25}} = 1 - \frac{20}{(0.01 \cdot 250 + 2)^2 (0.39 + 4)^{0.25}} = 0.318$, we obtain on the basis of (10.16) and Table 13: $K_{F\sigma} = K_{Fc} + (K^0_{F\sigma} - K_{Fc}) K_w = 0.34 + (0.741 - 0.34) 0.318 = 0.468$

Determining by [43] for the involute analogue $(K_{Fv})_{inv} = 1.003$, we obtain, according to (10.28), (10.29): $\mu = 4.23 \cdot 8^{-0.72} = 0.946$ and
$$(K_{Fv})_{nov.} = (1.003 - 1) 0.946 + 1 = 1.003.$$

Now we obtain the load coefficient $K_F = 1 \cdot 0.468 \cdot 1.003 = 0.469$.

With account of K_F we correct a^*_F (see above), multiplying it by $\sqrt[3]{K_F}$ and obtaining $a^*_F = 0.90$; then the new value $Y_{a2} = 0.945$.

Therefore: $\sigma_{F2} = 2000 \cdot 0.469 \cdot 30000 \cdot 0.59 \cdot 0.945 / (10^3 \cdot 70) = 224\ MPa < \sigma_{FP2} = 287\ MPa$.

For $n_k = 1$, assuming $\psi_{bd} = b_w / d_{w1} = 145/150 = 0.97$, according to (10.14) and Table 13 we obtain: $K^0_{F\sigma} = 1.427 \cdot 0.802^{0.436} \cdot 0.97^{0.302} \cdot 5^{-0.176} \cdot 10^{-0.020} = 0.924$ and $K_{F\sigma} = 0.34 + (0.924 - 0.34)0.318 = 0.526$

Then $K_F = 1 \cdot 0.526 \cdot 1.003 = 0.528$ and

$\sigma_{F2} = 2000 \cdot 0.528 \cdot 30000\, 0.59 \cdot 0.945/(10^3 \cdot 70) = 252\ MPa < \sigma_{FP2} = 287\ MPa$

Bending Strength under the Action of Maximum Load

According to [43], on the basis of the paragraph 13.2 and conclusions of Chapter 14, the allowable stress $\sigma_{FP\,max\,2} \approx \sigma^0_{FSt2} \cdot Y_{X2} / S_F = 6.5 \cdot 250 \cdot 0.96 / 1.1 = 1418\ MPa$.

Then, according to (13.11) for $n_k = 1$ we have: $\sigma_{F\,max\,2} = \sigma_{F2}(T_{max\,2} / T_2)^{0.8547} = 252(60000/30000)^{0.8547} = 456\ MPa < \sigma_{FP\,max\,2} = 1418\ MPa$

Contact Endurance

$$\sigma_{HPi} = \sigma_{H\,lim\,bi} Z_{Ni} Z_{vi} Z_{xi} / S_{Hi}.$$

1) Allowable stress [43]: $\sigma_{H\,lim\,b1} = 1050\ MPa$,

$\sigma_{H\,lim\,b2} = 2 \cdot HB_2 + 70 = 2 \cdot 250 + 70 = 570\ MPa$.

For $v = 0.39\ m/s$ we accept $Z_{v1} = Z_{v2} = 1$;

$Z_{x1} = 1,\ Z_{x2} = \sqrt{1.07 - 10^{-4} d_{w2}} = \sqrt{1.07 - 10^{-4} \cdot 750} = 0.997 \approx 1$;

$N_{H\,lim\,1} = 30(HB_1)^{2.4} = 30(550)^{2.4} = 113 \cdot 10^6$,

$N_{H\,lim\,2} = 30(HB_2)^{2.4} = 30(250)^{2.4} = 17 \cdot 10^6$;

$Z_{N1} = \sqrt[6]{N_{H\,lim\,1} / N_{H1}} = \sqrt[6]{113 \cdot 10^6 / 75 \cdot 10^6} = 1.07$,

$Z_{N2} = \sqrt[6]{N_{H\,lim\,2} / N_{H2}} = \sqrt[6]{17 \cdot 10^6 / 15 \cdot 10^6} = 1.02$.

According to conclusions of Chapter 14: $S_{H1} = S_{H2} = 1$. Then:

$\sigma_{HP1} = 1050 \cdot 1.07 \cdot 1 \cdot 1 = 1124\ MPa$;

$\sigma_{HP2} = 570 \cdot 1.02 \cdot 1 \cdot 1 = 581\ MPa$.

2) Acting stress. According to (11.8):

$\sigma_H = Z_H [K_H T_2 / (z \cos\alpha_k)]^{0.687} m^{-2.063} (\rho^*_\beta)^{-0.312} (Z_l K_l l^*)^{-1.063}$, here: Z_H = 3960, $l^* = 0.841$, $Z_l = 0.932$ and $K_l = 0.847$ (see above).

Letus define the necessary multipliers for the coefficient K_H (10.7).

We accept $K_A = 1$ (see above).

According to (10.19) $A_m = 0.36 \cdot 0.114^{0.16} 21^{0.6} = 1.58$, according to (10.18) for $n_\kappa = 0$: $K_{H\sigma m} = 0.5 \cdot 1.58 \cdot 0.468 = 0.370$, according to (10.20) $K^0_{H\sigma c} = 0.513 \cdot 0.034^{-0.117} \cdot 0.741 = 0.565$ (since $f_x = F_{\beta 1} = 0.034$ [40]).

According to the Table 13 we define $K_{Hc}=0.28$ and then (10.16): $K_{H\alpha c}=0.28+(0.565-0.28)0.318=0.371;$

According to (10.21): $K_{H\sigma}=max\{0.370,\ 0.371\}=0.371;$

Accepting the friction factor $f=0.07,$ we obtain (10.22): $K_f=(1+60\cdot 0.07^2 0.114^{0.25})^{0.728}=1.12;$

Considering, that according to the paragraph 10.6. $K_{Hv}< K_{Fv}$, we accept $K_{Hv}=1.$

Now, without account of the coefficient K_λ (we will describe it below) $K_H=1\cdot 0.371\cdot 1\cdot 1.12=0.416$ and then

$$\sigma_H=3960\left(\frac{30000\cdot 0.416}{70\cdot \cos 27^\circ}\right)^{0.687}\cdot 10^{-2.063}(107.17)^{-0.312}(0.932\cdot 0.847\cdot 0.841)^{-1.063}=467\ \ MPa.$$

For the case $n_k=1$ we have

$$K_{H\sigma m}=0.5\cdot 1.58\cdot 0.526=0.416,$$
$$K^0_{H\alpha c}=0.513\cdot 0.034^{-0.117}\cdot 0.924=0.705,$$
$$K_{H\alpha c}=0.28+(0.705-0.28)0.318=0.415,$$
$$K_{H\sigma}=max(0.416,\ 0.415)=0.416,$$
$$K_H=1\cdot 0.416\cdot 1\cdot 1.12=0.466,$$

$$\sigma_H=3960\left(\frac{30000\cdot 0.466}{70\cdot \cos 27^\circ}\right)^{0.687}\cdot 10^{-0.263}(107.17)^{-0.312}(0.932\cdot 0.847\cdot 0.841)^{-1.063}=505\ \ MPa.$$

Therefore, in both cases $\sigma_H<\sigma_{HP2}=581\ MPa$.

Adaptability of the Gearing

Adaptability of the gearing is characterized by the coefficient K_λ (see Chapter 10).

First, we determine the parameter Δr, accounting the radial errors of the gearing in the stochastic aspect.

The total tolerance for the shift of the basic rack profile (5.2) $\widetilde{T}_{H\Sigma}=0.685[0.06+0.012(9-3)]\cdot 10^{0.33}=0.193\ mm$ is divided between the pinion and gear (5.3): $\widetilde{T}_{H1}=0.193/(1+5^{0.2})=0.081mm,$

$\widetilde{T}_{H2}=0.193-0.081=0.112mm$.

Accepting [40] the radial run-out of gear rims for $k=9$ equal to $F_{r1}=0.112\ mm,$ $F_{r2}=0.125\ mm$ and the total tolerance f_a for the center distance 0.240 mm, we have the parameter (5.5) $R_c=0.112^2+0.125^2+0.081^2+0.112^2+0.24^2=0.105\ mm^2.$

Then, according to the Note in Chapter 5, we accept $K_c=2.4$ and on the basis of (5.4): $\Delta r=1.96\sqrt{0.105/2.4}=0.265\ mm$.

Calculating according to (11.27) the coefficient $n_b=0.549$, we determine the relative minor semi-axis of the contact ellipse: $b^*_H=a^*_H n_b/n_a=0.9\cdot 0.549/2.263=0.218$

Then we define the error, taken by the gearing with account of the pattern by the dimension b^*_H (Δa) and without account of the pattern, that is, for $b^*_H=0$ (Δa^0).

For this purpose, we will apply the relations (5.9) – (5.16):
$\Delta a_{ab} = 10 \cdot 0.12\{\sin[51.5^\circ - 0.218 \cdot 180^\circ/(1.15 \cdot \pi)] - \sin 27^\circ\} \cdot 0.847 = 0.201 mm,$

$\Delta j_w = j_w - j_{nmin} = 0.5 - 0.250 = 0.250\ mm,$

$\alpha_{aj} = \arccos[\cos 27^\circ - 0.25/(2 \cdot 10 \cdot 0.12)] = 38.11^\circ,$

$\Delta a_{aj} = 10 \cdot 0.12(\sin 38.11^\circ - \sin 27^\circ) = 0.196\ mm,$

$\Delta a_{ac} = mc^* - j_{nmin} = 10(1.05 - 0.9) - 0.25 = 1.25\ mm,$

$\Delta a_a = \min\{\Delta a_{ab}, \Delta a_{aj}, \Delta a_{ac}\} = 0.196\ mm;$

$\Delta a_p = 10 \cdot 0.12\{\sin 27^\circ - \sin[8.384^\circ + 0.218 \cdot 180^\circ/(1.15 \cdot \pi)]\} \cdot 0.847 = 0.126\ mm,$

$\Delta a = \Delta a_a + \Delta a_p = 0.196 + 0.126 = 0.322\ mm,$

$\Delta a^0_{ab} = 10 \cdot 0.12(\sin 51.5^\circ - \sin 27^\circ) \cdot 0.847 = 0.334\ mm,$

$\Delta a^0_{aj} = \Delta a_{aj} = 0.196 mm, \Delta a^0_{ac} = \Delta a_{ac} = 1.25 mm,$ $\Delta a^0_a = \min\{\Delta a^0_{ab}, \Delta a^0_{aj}, \Delta a^0_{ac}\} = 0.196 mm,$

$\Delta a^0_p = 10 \cdot 0.12(\sin 27^\circ - \sin 8.384^\circ) \cdot 0.847 = 0.313 mm,$

$\Delta a^0 = \Delta a^0_a + \Delta a^0_p = 0.196 + 0.313 = 0.509 mm$

Now, on the basis of (5.17) we obtain the parameter t: $t = (0.265 - 0.322)/(0.509 - 0.322) = -0.305.$

According to the paragraph 10.5 for $t < 0$ $(that is,\ \Delta r < \Delta a)$ we have $K_\lambda = 1$, that is, there is no "peak" of the stress at upper edges of the tooth and consequent reduction of the load-bearing capacity.

Contact Strength under the Action of Maximum Load

For the weakest element (driven gear) the allowable stress [43] is: $\sigma_{HP\max 2} = 2.8\sigma_T = 2.8 \cdot 400 = 1120\ MPa$. ($\sigma_T = 400\ MPa$ - yield point of the material).

The acting stress according to (11.29): $\sigma_{H\max} = 11050 Z_{HM}\{K_H T_{\max 2}(1 + C_{\alpha\beta})^2/[z_2 \cos\alpha_k(\rho^*_\alpha)^2]\}^{1/3}/(n_a n_b m).$

Assuming $\bar{\beta} = n_b/n_a = 0.549/2.263 = 0.243,$ we define by (11.21):

$Z_{HM} = \sqrt{(1 - 0.243 + 0.243^2)/(1 + 0.243)} = 0.727.$

Then for $n_k = 1$ (the least favorable case): $\sigma_{H\max} = \frac{11050 \cdot 0.727}{2.263 \cdot 0.549 \cdot 10}\sqrt[3]{\frac{60000 \cdot 0.466(1 + 0.114)^2}{70 \cos 27^\circ \cdot 12.17^2}}$
$= 1004 MPa < \sigma_{HP\max 2} = 1120 MPa.$

Evaluation of the Depth Contact Endurance

According to the paragraph 12.8 for nitrided gears, the strength condition for the depth contact (12.21) is: $\sigma_{k\max} \le (3.95...4.0)H_k$.

In our example the least $H_k = HV250$, then $\sigma_{k\max} \le 987 MPa$

Since, as it was noted in Chapter 12, nitrided gearing should be calculated as run-in, let us apply the formula (12.18) for definition of $\sigma_{k\max}$ and assuming (10.5), (10.7) we get

$$T_{H2} = 30000 \cdot 1 \cdot 0.416 \cdot 1 \cdot 1 \cdot 1 = 12480 N \cdot m,$$

$$\sigma_{k\,max} = 8626[12480/(70\cos 27°)]^{0.606} \cdot 10^{-1.82}(0.932 \cdot 0.847 \cdot 0.841)^{-0.821} \times$$
$$\times 10717^{-0.394} = 719 MPa$$

Therefore, the incipient depth crack in the pinion tooth core is excluded.

Example 2

It is necessary to perform the designing and also checking correcting analyses according to the DCS of the cylindrical Novikov gearing with the basic rack profile RGU-5, $m=3.15$ mm, $z_1 = 32, z_2 = 65, b_w = 45mm, \beta = 17.284°$ (see pos.10, 11 in the Table 21 and Fig. 31 in Chapter 12) for the following initial data: $\sigma_{k\,max} = 2174 MPa, b_H = 0.845mm, \bar{\beta} = 0.113,$ material – steel 25HGM, $H_0 = HV(700..750), h_0 = 0.20mm$

Letus perform the calculation for two versions of CHT – carburizing (version C) and nitro carburizing (version NC). Let us accept for the version C - $H_k = HV360,$ and for the version NC - $H_k = HV380.$ We assume, that in both cases the automatic control of the process of CHT is absent and the final machining after CHT is provided.

According to the instructions, given in the paragraph 12.8 of Chapter 12, we perform consequently the following actions.

1) With account of performing this calculation according to the initial (without run-in) geometry of contacting tooth flanks and taking into account explanations to formulas (12.7) and (12.8) from Chapter 12, we assign the coefficients:
   $Z_{LK} = 1.125, K_2 = 1, K_4 = 1.04, K_5 = 1, K_6 = 0.90.$
2) Definition of the total depth of the hardened layer.
   a) We calculate $(\sigma_{k\,max})_{con} = 2174/(1.125 \cdot 1 \cdot 1.04 \cdot 1 \cdot 0.90) = 2065 MPa$
   b) According to diagrams 3 and 4 of Fig. 27 we define $H_{kr} = HV640$ (for the version C) and $H_{kr} = HV580$ (for the version NC).
   c) According to the formula (12.6) and data from p. 2.2 we define $\chi_{kr} = 0.79$ (for the version C) and $\chi_{kr} = 0.87$ (for the version NC).
   d) According to the Table 19 we define $z_{kr}^0 = 0.75$ - for both versions.
   e) By means of the auxiliary diagram in Fig. 35 we determine $h_{te} = 0.90mm$ (for the version C) and $h_{te} = 0.60mm$ (for the version NC).
   f) According to the formula (12.13) we define $h_t = 1.85mm$ (for the version C) and $h_t = 1.30mm$ (for the version NC).
3) By means of formulas (12.11) and (12.12) we calculate the hardness distribution through the depth of the hardened layer with a certain step, the calculation results are convenient to be presented in the table:

z^0		0.1	0.3	0.5	0.7	0.9	1.1	1.3	1.5	1.75	2.0
$H^z(HV)$	C	714	729	719	685	616	536	459	400	365	360
$H^z(HV)$	NC	-	728	703	615	520	444	398	380	380	380

4) By means of the auxiliary Table 22 we determine the potentially "risky" area and fix its depth $z' = 1.75$ (for the version C) and $z' = 1.30$ (for the version NC).
5) According to the formula (12.1) and by means of the Table 22 we define $\sigma_e^0 = 0.425$ (for the version C) and $\sigma_e^0 = 0.501$ (for the version NC).
6) Now by (12.7) we calculate $\sigma_{HKPe}^0 = 0.416$ (for the version C) and $\sigma_{HKPe}^0 = 0.454$(for the version NC).
7) Then the required safety factor according to DCS will be determined as $S_{HK} = 0.98$ (for the version C) and $S_{HK} = 0.91$ (for the version NC).

The calculation showed, that if the chosen parameters of the diffusion layer can be considered satisfactory for the version C ($S_{HK} \approx 1$), then for the version NC these parameters are required to be corrected ($S_{HK} < 1$).

Thus, having increased the core hardness H_k up to $HV\ 400$, we get during the calculation of the version NC: $h_{te} = 0.65mm, h_t = 1.40mm, z' = 1.30, \sigma_e^0 = 0.501, \sigma_{HKPe}^0 = 0.492$, and then $S_{HK} = 0.98$, which can already be considered acceptable.

Therefore, the calculation showed, that:

a) it is possible to apply both carburizing and nitro carburizing for the considered gearing;
b) in both versions the "risky" area is at the sub-layer, and the level of DCS is determined, mainly, by the strength of the core, rather than of the effective layer.

Calculation results of several examples according to the developed computer program LSZPVK are given in Appendix 4.

APPENDIX 4

EXAMPLES OF THE COMPUTER PRINTOUT OF CYLINDRICAL NOVIKOV GEARING ANALYSIS RESULTS ACCORDING TO THE DEVELOPED CALCULATION PROGRAM LSZPVK

The program LSZPVK is intended for geometrical and strength analyses at personal computers of the cylindrical NG TLA, consisting of steel external gears with unground teeth, without strain hardening, operating under irreversible and reversible loads in continuous or intermittent (heavy-duty, medium-duty, light-duty) mode with lubrication in the closed casing at tangential speed under *25 m/s*.

Strength analysis is performed in the stochastic aspect with the chosen design value of the percent of "risk level", denoting, that there is the increased danger of the gearing failure within this value.

The program provides:

- calculation of main geometrical characteristics of the gearing with computation of quality factors of the engagement;
- calculation of the set of lengths of common normal lines and dimensions of the tooth for the edge gear tooth snap gauge with tolerances as applied to the batch production of gears;
- calculation of contact endurance of the teeth active flanks with account of their run-in and adaptability of the gearing;
- calculation of the tooth bending endurance.

The program covers helical and herring-bone Novikov gearing, made on the basis of one of the three most widespread basic rack profiles: RGU-5, according to the standard GOST 15023-76 and GOST 30224-96.

The types of heat treatment of the pinion and gear are nitro carburizing, martempering, normalizing and nitriding.

Longitudinal overlap ratio of teeth must be not less than *0.9* and under *4.5* (for helical gearing), under *2.25* for half of the herring-bone (for herring-bone gearing).

The degree of accuracy of the gearing according to kinematics, smoothness and contact rating is provided from 6 to 12 (according to the standard GOST 1643-81).

The pitch normal module of the gearing must be within the range *1.6...16 mm*.

The pitch diameter of the gear must be under *1600 mm*.

The given printouts show the initial data and results of calculation of geometry and load-bearing capacity (calculation of allowable gear torques) for the gearing, calculated manually in the Appendix 3 (Example 1).

The basic rack profile according to the standard GOST 15023-76 is designated by the word "GOST". The version 1 – for $n_k = 0$ (rigidity of shafts and bearing supports is not considered), the version 2 – for $n_k = 1$ (rigidity of shafts and supports is considered). The results show, that in the second case the load-bearing capacity of the gearing is lower, than in the first case according to both contact and bending endurance.

Version N1 **Program LSZPVK**

INITIAL DATA GEARING TYPE: CYLINDRICAL HELICAL NOVIKOV GEARING

Basic rack profile	Standard GOST	Type of heat treatment of the pinion	Nitriding
Nominal center distance, mm	450.00	Type of heat treatment of the gear	Martempering
Normal pitch module, mm	10.00		
Number of pinion teeth	14	Hardness of the pinion tooth according to Brinell	300.0
Number of gear teeth	70		
Operating width of toothing (half of the herring-bone), mm	150.00	Hardness of the gear tooth according to Brinell	250.0
Helix angle of the rack, deg.	21.000	Rotational frequency of the pinion, 1/min	50.0
Shift coefficient of the pinion	0.082		
Degree of accuracy according to kinematics	9	Design time of the pair operation, thous. hours	25.000
Degree of accuracy according to smoothness	8	Feature of the gearbox design	0
Degree of accuracy according to contact	8	Loading mode Design value of the "risk level", %	RP 0.5 5.000

Calculation Results
1. Geometry

Addendum diameter of pinion teeth, mm	169.60	LIMITING FACTORS	
Addendum diameter of gear teeth, mm	766.40	Maximum allowable value of the pitch point shift	0.131
Root diameter of pinion teeth, mm	130.60	Actual value of the pitch point shift	0.080
Root diameter of gear teeth, mm	727.40	Angle of undercut of the pinion addendum, deg.	7.040
Nominal tooth height, mm	19.50		
Coefficient of the sum of shifts	0.012	Angle of undercut of the gear addendum, deg.	1.999
Shift coefficient of the gear	-0.070	Pinion tooth thickness at the tip	0.629
Axial overlap ratio (for half of the herring-bone)	1.711	Gear tooth thickness at the tip	0.646

2. Reference dimensions of the tooth

LENGTH OF THE COMMON NORMAL, mm						TOOTH DIMENSIONS FOR THE EDGE GEAR TOOTH SNAP GAUGE, mm					
PINION			GEAR			PINION			GEAR		
						Height	Thickness		Height	Thickness	
77.528	0.025 -0.021	(1)	292.701	0.040 -0.032	(1)	Add.: 3.126	11.723	0.048 -0.040	Add.: 3.645	12.492	0.064 -0.052
106.450	0.043 -0.035	(1)	322.229	0.045 -0.036	(1)	Root: 14.487	16.757	0.044 -0.036	Root: 14.456	17.998	0.066 -0.053
132.152	0.059 -0.048	(1)	351.327	0.049 -0.039	(1)						
0.000	0.000 0.000	(0)	379.945	0.053 -0.043	(1)						

3. Load-bearing capacity of gearing

DESIGN (for variable modes – maximum according to the cyclogram)
ALLOWABLE TORQUE AT THE GEAR, Nm
According to surface contact endurance – 47911.7 (47911.7)
According to bending endurance – 41180.9

Version № **Program LSZPVK**

INITIAL DATA GEARING TYPE: CYLINDRICAL HELICAL NOVIKOV GEARING

Basic rack profile	Standard GOST	Type of heat treatment of the pinion	nitriding
Nominal center distance, mm	450.00	Type of heat treatment of the gear	martempering
Normal pitch module, mm	10.00	Hardness of the pinion tooth according to Brinell	300.0
Number of pinion teeth	14	Hardness of the gear tooth according to Brinell	250.0
Number of gear teeth	70	Rotational frequency of the pinion, 1/min	50.0
Operating width of toothing (half of the herring-bone), mm	150.00	Design time of the pair operation, thous. hours	25.000
Helix angle of the rack, deg.	21.000	Feature of the gearbox design	13
Shift coefficient of the pinion	0.082	Loading mode	RP 0.5
Degree of accuracy according to kinematics	9	Design value of the "risk level", %	5.000
Degree of accuracy according to smoothness	8		
Degree of accuracy according to contact	8		

Calculation Results

1. Geometry

		LIMITING FACTORS	
Addendum diameter of pinion teeth, mm	169.60	Maximum allowable value of the pitch point shift	0.131
Addendum diameter of gear teeth, mm	766.40	Actual value of the pitch point shift	0.080
Root diameter of pinion teeth, mm	130.60	Angle of undercut of the pinion addendum, deg.	7.040
Root diameter of gear teeth, mm	727.40	Angle of undercut of the gear addendum, deg.	1.999
Nominal tooth height, mm	19.50	Pinion tooth thickness at the tip	0.629
Coefficient of the sum of shifts	0.012	Gear tooth thickness at the tip	0.646
Shift coefficient of the gear	-0.070		
Axial overlap ratio (for half of the herring-bone)	1.711		

2. Reference dimensions of the tooth

<table>
<tr><th colspan="6">LENGTH OF THE COMMON NORMAL, mm</th><th colspan="6">TOOTH DIMENSIONS FOR THE EDGE GEAR TOOTH SNAP GAUGE, mm</th></tr>
<tr><th colspan="3" rowspan="2">PINION</th><th colspan="3" rowspan="2">GEAR</th><th colspan="3">PINION</th><th colspan="3">GEAR</th></tr>
<tr><th>Height</th><th colspan="2">Thickness</th><th>Height</th><th colspan="2">Thickness</th></tr>
<tr><td>77.528</td><td>0.025
-0.021</td><td>(1)</td><td>292.701</td><td>0.040
-0.032</td><td>(1)</td><td>Add.:
3.126</td><td>11.723</td><td>0.048
-0.040</td><td>Add.:
3.645</td><td>12.492</td><td>0.064
-0.052</td></tr>
<tr><td>106.450</td><td>0.043
-0.035</td><td>(1)</td><td>322.229</td><td>0.045
-0.036</td><td>(1)</td><td>Root:
14.487</td><td>16.757</td><td>0.044
-0.036</td><td>Root:
14.456</td><td>17.998</td><td>0.066
-0.053</td></tr>
<tr><td>132.152</td><td>0.059
-0.048</td><td>(1)</td><td>351.327</td><td>0.049
-0.039</td><td>(1)</td><td colspan="6" rowspan="2"></td></tr>
<tr><td>0.000</td><td>0.000
0.000</td><td>(0)</td><td>379.945</td><td>0.053
-0.043</td><td>(1)</td></tr>
</table>

3. Load-bearing capacity of gearing

DESIGN (for variable modes – maximum according to the cyclogram)
ALLOWABLE TORQUE AT THE GEAR, Nm
According to surface contact endurance – 40367.0 (40367.0)
According to bending endurance – 34421.9

The allowable design torque at the gear according to the surface contact endurance is pointed in brackets, which can be transmitted by the gearing for the given degree of kinematic accuracy with the contact pattern at the tooth edge.

The printouts show, that in both cases the results in brackets and without brackets coincide, which indicates the sufficient adaptability of the gearing (there is no reduction of the load-bearing capacity).

The program LSZPVK allows to design easily and effectively the cylindrical Novikov gearing with optimal parameters, providing the increased load-bearing capacity, the reduced material consumption and the high competitiveness of the drive in the world market.

The Laboratory of Special Gearing of VMAMRI SFU is ready to provide customers with the computational program LSZPVK on the contract basis and render its assistance in application of this program, for this purpose it is necessary to make a request to the address, pointed in the Foreword to this monograph.

APPENDIX 5

EXAMPLES OF THE COMPUTER PRINTOUT OF BEVEL NOVIKOV GEARING ANALYSIS RESULTS ACCORDING TO THE DEVELOPED CALCULATION PROGRAM CONUS

The program CONUS is intended for calculations of bending and contact stresses of bevel Novikov gearing teeth, operating at average tangential speed under *25 m/s*.

Strength analysis is performed in the stochastic aspect with the chosen design value of the percent of "risk level", denoting, that there is the increased danger of the gearing failure within this value.

The analysis provides the absence or presence of the longitudinal tooth flanks ease-off in the gearing, that is, their modification (localization), applied in practice for increasing the gearing working capacity. In case of localization it is assumed, that the surface ease-off is performed due to the increase of cutter head diameter for cutting the pinion (compared with the diameter of the cutter head, determined according to the condition of a rigid non-congruent pair) when manufacturing the pair by a simple two-sided method.

The program CONUS provides the computation of bending (separately for the pinion and the gear) and contact stresses for the assigned calculated number of engagement phases with the consequent choice of maximum stresses for all phases.

The program covers orthogonal bevel Novikov gearing, made on the basis of one of the three most widespread basic rack profiles: RGU-5, according to the standard GOST 15023-76 and GOST 30224-96.

Gears teeth must be made of equal height along the whole width of toothing (axial shape III according to the standard GOST 19326-73).

The pinion toothing width is equal to the gear toothing width.

The axial overlap ratio is within the range *0.9...2.8*.

Degrees of accuracy of gears according to smoothness and contact ratings are provided from 6 to 12 (according to standards GOST 1758-81 and GOST 9368-81).

Module of the tool must be within the range *1.6...16 mm*.

The mean pitch diameter of the gear must be under *1600 mm*.

The structural design of the gearing provides the case, widespread in practice:

- the pinion-shaft and the gear are mounted on two supports;
- the mounting of the pinion-shaft is cantilever;

- the diameter of the pinion-shaft is close to the inner diameter of the pinion at the external face;
- the distance between supports of the pinion-shaft is approximately equal to the double distance from the leading support to the middle of the pinion gear rim;
- the torque input is from the tail end (that is, behind the end support) of the pinion-shaft;
- the pinion is made with the left tooth inclination and with the concave flank, and the gear – with the right tooth inclination and with the convex flank, which provides the action of axial loads towards the gears supports.

The number of calculated engagement phases (segments of partitioning of the toothing width) is assigned by an integer number from *1* and more. Each phase is calculated for the pinion and the gear for different directions of the vector of gears manufacturing errors, that is, both for increasing of kinematic clearances from the internal face to the external one (the reference point is at the internal face), and otherwise (the reference point is at the external face).

As the main calculation result a table is printed for maximum bending stresses for the pinion and the gear and maximum contact stresses with data, characterizing the engagement phase, design point and error vector (position of the reference point), for which the pointed maximum stresses have been obtained. Coordinates of points are given in parts of the toothing width, the "minus" sign here is from the middle to the inner face, the "plus" sign - to the outer face.

Printouts of the calculation results are given below for two examples, differing by the absence (example 1) and presence (example 2) of the longitudinal localization of tooth flanks. In both cases the design number of engagement phases is taken to be equal to *21*.

Examples show, that the theoretical number of contact points varies from *1* to *3*.

The obtained results for the example 1 are:

a) the maximum bending stress at the pinion tooth (*415.1 MPa*) is in the section, where the pinion addendum is contacting with the gear dedendum at the point with the coordinate *–0.454*, the engagement phase here corresponds to the position of the gear addendum (pinion dedendum) at the point with the coordinate *0.050*; the vector of clearances is directed from the inner face end to the outer one (the reference points is at the inner face end);
b) the maximum bending stress at the gear tooth (*461.5 MPa*) is in the section, where the gear addendum is at the outer face end (the coordinate *0.500*), the engagement phase here corresponds to the position of the gear addendum also at the outer face end (the coordinate *0.500*), but with the reverse direction of the vector of clearances compared with the case *a*);
c) the maximum contact stress (*1172.1 MPa*) coincides with the case *b*) according to the position of the calculation point, the engagement phase and also the direction of the vector of clearances.

As for the example 2, having the value of longitudinal ease-off of pinion tooth flanks at face ends $\delta l_T = 0.07$ *mm*, maximum bending stresses, as it is seen in printouts, for the pinion is *415.2 MPa*, for the gear – *371.1 MPa*, the maximum contact stress - *968.6 MPa*.

Comparison of results of two examples indicates, that in both cases the acting maximum bending stresses are far from the allowable value for high-hardened teeth ($\sigma_{FP} \approx 650...750$ *MPa*). Maximum contact stress for the example 1 is almost equal to the allowable one (σ_{HP}= *1200 MPa*), therefore, the load T_2 $=2000$ $N \cdot m$ in this case is close to the allowable one. For the second example, the maximum contact stress is *1.21* times less than for the example 1 and, therefore, it is considerably less than the allowable stress. This happened because the corresponding contact point is significantly removed from the face (the coordinate *0.163*). Calculations show, that the allowable contact stress for the example 2 is achieved at the load $T_2 = 2800$ $N \cdot m$, the maximum bending stresses still remaining here considerably less than the allowable ones.

Therefore, the comparison of calculation results of the two examples shows in this case the benefit from the longitudinal modification (localization) of tooth flanks, represented in *1.4* times increase of the gearing load-bearing capacity.

The program CONUS allows to design the bevel Novikov gearing with optimal parameters of longitudinal modification of teeth, providing the increased load-bearing capacity and the reduced material consumption of the drive.

The Laboratory of Special Gearing of VMAMRI SFU is ready to provide customers with the computational program CONUS on the contract basis and render its assistance in application of this program, for this purpose it is necessary to make a request to the address, pointed in the Foreword to this monograph.

Example 1 **Program CONUS**

CALCULATION OF STRESSES IN ORTHOGONAL BEVEL NOVIKOV GEARING

INITIAL DATA

BASIC RACK PROFILE	Standard GOST
Tool module, mm	6.000
Normal mean module of the gearing, mm	6.000
Number of pinion teeth	10
Number of gear teeth	31
Mean helix angle of the tooth mean line, deg.	35.000
Shift coefficient of the pinion	0.000
Coefficient of variation of the tooth design thickness of the pinion	0.000
Coefficient of the toothing width	0.226
Nominal diameter of the cutter head, mm	177.696
Degree of accuracy according to smoothness rating (GOST 1758-81 and GOST 9368-81)	10
Degree of accuracy according to contact rating (GOST 1758-81 and GOST 9368-81)	10
Torque at the gear, Nm	2000.00
Rotational frequency of the pinion, 1/min	1500.00
Minimum pinion and gear tooth hardness according to Brinell	560
Feature of the influence of the pinion-shaft torsion	1
Feature of the influence of the pinion-shaft deflection	1
Longitudinal (symmetrical) ease-off at face surfaces for localization, mm	0.000
Coordinate of the localization center in parts of the toothing width	0.000
Design "risk level", %	3.000
Number of calculated engagement phases (within the toothing width)	21

Calculation Results

1. Geometry

Mean cone distance, mm	119.293
Operating width of toothing, mm	30.395
Mean axial overlap ratio	0.925
Minimum theoretical number of contact points for all phases	1
Maximum theoretical number of contact points for all phases	3

2. Table of maximum stresses

Stress, MPa		Engagement phase				Design point			Position of the reference point (without localization)
		Number of contact points		Addendum of the pair element	Coor-dinate	Number of tooth	Addendum of the pair element	Coor-dinate	
		Theor.	Actual						
σ_{F1}	415.1	2	2	Gear	0.050	3	Pinion	-0.454	At the inner face surface
σ_{F2}	461.5	2	2	Gear	0.500	4	Gear	0.500	At the outer face surface
σ_H	1172	2	2	Gear	0.500	4	Gear	0.500	At the outer face surface

Example 2 **Program CONUS**

CALCULATION OF STRESSES IN ORTHOGONAL BEVEL NOVIKOV GEARING INITIAL DATA

BASIC RACK PROFILE	Standard GOST
Tool module, mm	6.000
Normal mean module of the gearing, mm	6.000
Number of pinion teeth	10
Number of gear teeth	31
Mean helix angle of the tooth mean line, deg.	35.000
Shift coefficient of the pinion	0.000
Coefficient of variation of the tooth design thickness of the pinion	0.000
Coefficient of the toothing width	0.226
Nominal diameter of the cutter head, mm	177.696
Degree of accuracy according to smoothness rating (GOST 1758-81 and GOST 9368-81)	10
Degree of accuracy according to contact rating (GOST 1758-81 and GOST 9368-81)	10
Torque at the gear, Nm	2000.00
Rotational frequency of the pinion, 1/min	1500.00
Minimum pinion and gear tooth hardness according to Brinell	560
Feature of the influence of the pinion-shaft torsion	1
Feature of the influence of the pinion-shaft deflection	1
Longitudinal (symmetrical) ease-off at face surfaces for localization, mm	0.070
Coordinate of the localization center in parts of the toothing width	0.000
Design “risk level”, %	3.000
Number of calculated engagement phases (within the toothing width)	21

Calculation Results

1. Geometry

Mean cone distance, mm	119.293
Operating width of toothing, mm	30.395
Mean axial overlap ratio	0.925
Minimum theoretical number of contact points for all phases	1
Maximum theoretical number of contact points for all phases	3

2. Table of maximum stresses

Stress, MPa		Engagement phase				Design point			Position of the reference point (without localization)
		Number of contact points		Addendum of the pair element	Coord-inate	Number of tooth	Addendum of the pair element	Coord-inate	
		Theor.	Actual						
σ_{F1}	415.2	2	1	Gear	-0.500	4	Pinion	0.163	At the outer face surface
σ_{F2}	371.1	2	1	Pinion	-0.350	5	Gear	0.163	At the outer face surface
σ_H	968.6	2	1	Gear	-0.500	4	Pinion	0.163	At the outer face surface

BIBLIOGRAPHY

[1] Aivazyan S.A. *Statistic investigation of relationships*. Moscow: Metallurgiya. 1968. 268 p. (in Russian).

[2] *Analysis of elastic interaction of tooth surfaces of gearing, formed by spiral-disk tools* / Anosova T.P., Erikhov M.L., Syzrantsev V.N., Sheveleva G.I. // Mashinovedenie. 1984. N 3. P. 45-50 (in Russian).

[3] Arefyev I.I. *Investigation of load-bearing capacity of Novikov gearing with two lines of action* // Toothed and worm gearing / Ed. by N.I. Kolchin. Leningrad: Mashinostroenie. 1968. P. 235-253 (in Russian).

[4] Akhtyrets G.P., Korotkin V.I. *Application of FEM to solution of contact task of elasticity theory with variable contact area* // Izvestiya SKNTs VSh. Estestvennye nauki. Rostov-on-Don: RSU Publishers. 1984. N 1, P. 38-42 (in Russian).

[5] Akhtyrets G.P., Korotkin V.I. *On solution of contact task by means of finite element method // Continuum mechanics.* Rostov-on-Don: RSU Publishers. 1988. P. 43-48 (in Russian).

[6] Akhtyrets G.P., Korotkin V.I., Pavlenko A.V. *Determination and evaluation of stress-strain state under action of concentrated force on Novikov gear tooth // Machine parts.* Republic interdepartmental scientific-technical digest. Kiev: Tekhnika. 1987. P. 49-55 (in Russian).

[7] Badaev A.M. *Principles of control and design of measuring devices for checking helix angle of circular teeth of bevel Novikov gearing* // Izvestiya vuzov. Mashinostroenie. 1968. N 4. P. 30-35 (in Russian).

[8] Baindurov V.S., Pavlenko A.V. *Investigation of bending stress in teeth of cylindrical gears* // Nadezhnost i kachestvo zubchatykh peredach. Moscow: NIINFORMTYaZhMASh. 1967. N 18-67-66. 9 p. (in Russian).

[9] Balakshin B.S. *Dimension chains* // Encyclopedic reference book. Vol.5. Section 3. Moscow: Mashinostroenie. 1947. P. 104-106 (in Russian).

[10] Bondarenko V.S., Kirichenko L.A., Kuzovkov B.P. *Technical economic effectiveness of long-term operation of Novikov gearing in a gearbox of turbo-geared aggregate* // Results of investigation and practical application of M.L. Novikov gearing. Reports theses of republic scientific-technical conference. Kharkov: 1971. P. 157 (in Russian).

[11] Bragin V.V. *Determination of tooth stress-strain state in cylindrical gearing* // Izvestiya vuzov. Mashinostroenie. 1981. N 9. P. 31-34 (in Russian).

[12] Brostsait E., Tsvirlyain O. *Internal stresses and their influence on stressed state of material at Hertz contact.* // Problemy treniya i smazki. 1986. N 3. P. 64–70 (in Russian).

[13] Budyka Yu.N. *Theory of engagement and comparative wear resistance of plane general type gearing* / Proceedings of seminar on theory of machines and mechanisms. Vol. 10. Issue 39. Moscow: AN SSSR. 1951. P. 56-74 (in Russian).

[14] Burov O.F. *Rupture and wear of teeth of double-point contact Novikov gearing with different basic rack profiles* // Results of investigation and practical application of M.L. Novikov gearing. Reports theses of republic scientific-technical conference. Kharkov: 1971. P. 108-112 (in Russian).

[15] Winter Kh., Weiss T. *Some factors, influencing fatigue pitting, micro-pitting and low speed wear of surface hardened gears* // Konstruirovanie i tekhnologiya mashinostroeniya. 1981. N 2. P. 135–142 (in Russian).

[16] Vasilyev V.M. *Analytical investigation of teeth undercut in three-dimensional engagements* / Proceedings of NPI. Vol.149. Novocherkassk: 1963. P. 40-48 (in Russian).

[17] Vasilchenko N.M. *Investigation of plastic bevel Novikov gearing* // Author's abstract of Ph.D. thesis. Voroshilovgrad: 1970. 24 p. (in Russian).

[18] Vasin G.M. *Investigation of non-zero cylindrical Novikov gearing* // Author's abstract of Ph.D. thesis. Moscow: 1972. 20 p. (in Russian).

[19] Vasin G.M., Mochalov V.L. *On tooth thinning of Novikov gears, manufactured with shift of basic rack profile* // Mechanical gearing. Izhevsk: "Udmurtiya" Publishers. 1972. P. 168-172 (in Russian).

[20] Veretennikov V.Ya. *Investigation of load-bearing capacity of general-purpose gearboxes* // Investigation and analysis of mechanical gearing. Izhevsk: "Udmurtiya" Publishers. 1966. P. 69-81 (in Russian).

[21] Veretennikov V.Ya., Koloshko V.P. *Investigation of gearboxes with double-point contact Novikov gearing under variable operation modes* // Mechanical gearing. Izhevsk: "Udmurtiya" Publishers. 1967. P. 17-21 (in Russian).

[22] Veretennikov V.Ya., Korotkin V.I., Boldyreva L.G. *Investigation of nitro carburized cylindrical Novikov gearing for general-purpose gearboxes* // Vestnik mashinostroeniya. 1984. N 6. P. 21-23 (in Russian).

[23] Veretennikov V.Ya. *Experience of work in laboratory of gearboxes testing.* Moscow: NIIMASh. 1971. P. 90-93 (in Russian).

[24] *Influence of faces closeness on teeth compliance and load distribution in real cylindrical TLA Novikov gearing* / Akhtyrets G.P., Katsevich A.Ya., Korotkin V.I., Pavlenko A.V., Fomenko V.E. / Reports theses of scientific-technical conference "Standardization and unification in the area of gearing". Kharkov: 1990. P. 33-34 (in Russian).

[25] Vodopyanov A.F., Zhukov V.A., Logvinenko E.N. *Experience of Novikov gearing application in gearboxes of centrifugal machines // Theory and geometry of three-dimensional engagement.* Reports theses of the III all-USSR symposium. Kurgan: 1979. P. 84-86 (in Russian).

[26] Vodopyanov A.F. *Investigation of sensitivity of cylindrical gearing with point contact to mounting errors* // Author's abstract of Ph.D. thesis. 1974. 24 p. (in Russian).

[27] Volchinsky A.G. *Investigation of load-carrying capacity of bevel TLA Novikov gearing with increased module* // Ph.D. thesis. Moscow: 1982. 141 p. (in Russian).

[28] Volchinsky A.G., Chesnokov V.A. *Increasing the load-carrying capacity of bevel gearing* // Izvestiya vuzov. Mashinostroenie. 1977. N 4. P. 40-44 (in Russian).

[29] Galanov B.A. *Method of boundary equations of Gammershtein type for contact problems of elasticity theory in case of unknown contact areas* // PMM. 1985. Vol. 49. Issue 5. P. 827-835 (in Russian).

[30] Galin L.A., Goryacheva I.G. *Three-dimensional contact problem on motion of die with friction* // PMM. 1982. Vol. 46. Issue 6. P. 1016-1022 (in Russian).

[31] Galper R.R. *Depth contact strength of hardened flanks // Increasing the load-carrying capacity of mechanical drive* / Ed. by V.N. Kudryavtsev. Leningrad: Mashinostroenie. 1973. P. 102-113 (in Russian).

[32] Galper R.R. *Contact strength of gearing with surface hardening*. Leningrad: LDNTP. 1964. 28 p. (in Russian).

[33] Geller Yu.A. *Tool steels*. Moscow: Mashinostroenie. 1983. 525 p. (in Russian).

[34] Genkin M.D., Krasnoshchekov N.N., Chesnokov V.A. *Experience of Novikov gearing implementation* // Investigation and development of Novikov gearing. Moscow: AN SSSR. 1960. P. 149-154 (in Russian).

[35] Ginzburg E.G. *Theoretical and experimental investigation of a variant of bevel Novikov gearing, cut by generation* // Author's abstract of Ph.D. thesis. Leningrad: 1960. 18 p. (in Russian).

[36] State standard GOST 11902-77. *Cutter heads for bevel and hypoid gears with circular teeth*. Moscow: Publishers of standards. 1985. 10 p. (in Russian).

[37] State standard GOST 14186-69. *Novikov cylindrical gears. Modules*. Moscow: Publishers of standards. 1969. 1 p. (in Russian).

[38] State standard GOST 19326-73. *Spiral bevel gearing. Calculation of geometry*. Moscow: Publishers of standards. 1974. 75 p. (in Russian).

[39] State standard GOST 30224-96. *Novikov cylindrical gearing with tooth surface hardness not less than 35 HRC_E. Basic rack profile.* // Intergovernmental standard. Intergovernmental Soviet on standardization, metrology and certification. 1996. Minsk. 5 p. (in Russian).

[40] State standard GOST 1643-81. *Cylindrical gearing. Tolerances*. Moscow: Publishers of standards. 1981. 69 p. (in Russian).

[41] State standard GOST 2185-66. *Cylindrical gearing. Basic parameters*. Moscow: Publishers of standards. 1966. 5 p. (in Russian).

[42] State standard GOST 16532-70. *Cylindrical involute external gearing. Calculation of geometry*. Moscow: Publishers of standards. 1970. 9 p. (in Russian).

[43] State standard GOST 21354-87. *Cylindrical involute gearing. Strength analysis*. Moscow: Publishers of standards. 1988. 125 p. (in Russian).

[44] State standard GOST 17744-72. *Novikov cylindrical gearing with two lines of action. Calculation of geometry*. Moscow: Publishers of standards. 1972. 16 p. (in Russian).

[45] State standard GOST 15023-76. *Cylindrical Novikov gearing with two lines of action. Basic rack profile*. Moscow: Publishers of standards. 1976. 3 p. (in Russian).

[46] State standard GOST 16771-81. *Finishing single-thread hobs for cylindrical Novikov gearing with two lines of action. Technical requirements*. Moscow: Publishers of standards. 1981. 21 p. (in Russian).

[47] Grebenyuk A.I. *Investigation of some problems of Novikov gearing teeth profiling //* Author's abstract of Ph.D. thesis. Kharkov: 1969. 20 p. (in Russian).

[48] Gribanov V.M. *Analytical theory of accuracy of three-dimensional engagement //* Izvestiya vuzov. Mashinostroenie. 1982. N 4. P. 49-52 (in Russian).

[49] Gribanov V.M. *Local kinematics of Novikov gearing under presence of manufacturing and assembly errors //* Izvestiya vuzov. Mashinostroenie. 1984. N 6. P. 51-54 (in Russian).

[50] Gribanov V.M., Shved O.P. *On accuracy of cylindrical Novikov gearing with high hardness of tooth flanks //* Mashinovedenie. 1984. N 3. P. 51-53 (in Russian).

[51] Grishko V.A. *Increasing of gearing wear resistance*. Moscow: Mashinostroenie. 1977. 232 p. (in Russian).

[52] Gulyaev K.I., Livshits G.A. *Synthesis and character of engagement of approximate gearing. // Theory and geometry of three-dimensional engagement.* Reports theses of the Second symposium. Leningrad: LDNTP. 1973. P. 9-10 (in Russian).

[53] Gulyaev K.I. *Approximate engagement synthesis by changeover points //* Toothed and worm gearing / Ed. by N.I. Kolchin. Leningrad: 1974. P. 17-23 (in Russian).

[54] Gutman E.A. *Novikov gearing in gearboxes of mining conveyors //* Novikov gearing. Issue I. Moscow: TsINTIAM. 1964. P. 284-291 (in Russian).

[55] Davydov Ya.S. *Some questions of Novikov gearing geometry //* Novikov gearing. Issue 2. Moscow: ZAFEA. 1962. P. 49-64 (in Russian).

[56] Derbasov N.M., Todorov V.S., Shishov V.P. *Investigation and implementation of Novikov gearing in coal-cleaning facilities // Results of investigation and practical application of M.L.* Novikov gearing. Reports theses of republic scientific-technical conference. Kharkov: 1971. P. 132 (in Russian).

[57] Dergausov A.U., Zhuravlev G.A., Ivanov B.D. *Manual on geometrical, strength and setting calculations of spiral bevel Novikov gearing during cutting on machine-tools of models 525, 528S and 5A284 and cylindrical Novikov gearing.* Saratov: TsNTI. 1969. 134 p. (in Russian).

[58] Dergausov A.U. *Results of implementation of Novikov gearing with high tooth hardness* // Vestnik mashinostroeniya. 1983. N 6. P. 30-32 (in Russian).

[59] *Double-point contact Novikov gearing: Inventor's certificate USSR 735852. B.I. 1980. N 19. (in Russian).*

[60] Dreizen I.S., Murygin O.P. *Experience of Novikov gearing implementation at Rostov machine-building plant //* M.L. Novikov gearing. Proceedings of NIITM. Issue X. Rostov-on-Don: 1964. P. 241-253 (in Russian).

[61] Drozd M.S., Tesker E.I. *Conditions of depth contact failure formation in steels with case hardened and nitro carburized layer for linear initial contact //* Mashinovedenie. 1976. N 1. P. 81-87 (in Russian).

[62] Dusev I.I. *Investigation of contact nature in approximate three-dimensional engagement //* Three-dimensional and hypoid gearing. Proceedings of NPI. Novocherkassk: 1970. Vol. 213. P. 33-41 (in Russian).

[63] Dusev I.I. *Undercut of gears teeth during their cutting //* Izvestiya vuzov. Mashinostroenie. 1965. N 6. P. 28-35 (in Russian).

[64] Erikhov M.L. *On question of synthesis of engagement with point touching //* Theory of gearing in machines. Moscow: Mashinostroenie. 1966. P. 78-91 (in Russian).

[65] Zablonsky K.I. *Gearing. Load distribution in engagement*. Kiev: Tekhnika. 1977. 208 p. (in Russian).

[66] Zinchenko V.M. *Engineering of gears surface by chemically heat treatment methods*. Moscow: Publishers of MGTU im. N.E. Baumana. 2001. 302 p. (in Russian).

[67] Zubarev N.I., Igdalov M.P. *Optimization of quality parameters of the gearing*. // Tractory i selkhozmashiny. 1989. N 2. P. 41-69 (in Russian).

[68] *Gearing: Inventor's certificate of USSR 675234. B.I. 1979. N 27 (in Russian).*

[69] *Gearing: Inventor's certificate of USSR 735855. B.I. 1980. N 19 (in Russian).*

[70] *Gearing for large torques: Patent USA 3982445. B.I. 1977. N 3*

[71] *TLA Novikov gearing: Inventor's certificate of USSR 1700319. B.I. 1991. N 4 (in Russian).*

[72] *Novikov gearing: Inventor's certificate of USSR 1013655. B.I. 1983. N 15 (in Russian).*

[73] *Novikov gearing: Inventor's certificate of USSR 796562. B.I. 1981. N 2 (in Russian).*

[74] *Gearing. Reference book* / Ed. by E.G. Ginzburg. Leningrad: Mashinostroenie. 1980. 416 p. (in Russian).

[75] Ivanov B.D. *Manual on geometrical and setting calculation for cutting hypoid Novikov pairs on machine-tools of models 525, 528S and 528SP*. Saratov: Publishers of Privolzhsky TsBTI. 1967. 56 p. (in Russian).

[76] Ivanov B.D., Poltavsky O.A. *Finishing tool profile determination during form-relieving of teeth cutters with any type of basic rack profile // Final, finish-strengthening and shape generating metal machining*. Proceedings of RISHM. Rostov-on-Don: 1973. P. 195-202 (in Russian).

[77] *Manufacturing of cutters for bevel and hypoid Novikov gears on machine-tools SZTZS*. Saratov: TsNTI. 1972. 5 p. (in Russian).

[78] Ioannidis, Kharris. *New model of rolling bearings durability*. // Friction and lubrication problems. 1985. N 3. P. 44–58 (in Russian).

[79] Isibasi, Iosino. *Comparison of efficiency and friction coefficient in involute and Novikov gearing // Konstruirovanie i tekhnologiya mashinostroeniya*. 1985. N 1. P. 228-237 (in Russian).

[80] Ismail-Zade A.I. *Experimental investigation of Novikov gearing gearboxes operation at variable loads* // Novikov gearing. Issue 2. Moscow: ZAFEA. 1964. P. 82-92 (in Russian).

[81] *Investigation and implementation of Novikov gearing in general-purpose gearboxes* / Korotkin V.I., Kharitonov Yu.D., Veretennikov V.Ya., Pavlenko A.V., Akhtyrets G.P. // Vestnik mashinostroeniya. 1990. N 12. P. 57-60 (in Russian).

[82] *Investigation and development of Novikov gearing*. Moscow: AN SSSR. 1960. 195 p. (in Russian).

[83] *Investigation of lubrication conditions of involute and Novikov gearing* // Ekspress-informatsiya. Detali mashin. 1987. N 33. P. 16-25 (in Russian).

[84] *Basic rack profile of double Novikov gearing teeth*: Inventor's certificate of USSR 240428. B.I. 1969. N 12 (in Russian).

[85] *Basic rack profile, accuracy standards and tooth machining technology of Novikov gearing with two lines of action* / Likhtsier M.B., Arkhangelsky L.A., Kogan G.I., Kopf I.A. Moscow: TsNIITMASh. 1966. 186 p. (in Russian).

[86] Itkis M.Ya. *Geometrical calculation of cylindrical Novikov gearing*. Volgograd: Nizhne-Volgskoe knizhnoe Publishers. 1973. 312 p. (in Russian).

[87] Karlash G.P. *Flanking of Novikov gears with one basic rack profile* // M.L. Novikov gearing. Proceedings of NIITM. Issue X. Rostov-on-Don: 1964. P. 208-213. (in Russian).

[88] Katsevich A.Ya., Kopasenko V.V. *Solution of some three-dimensional problems of elasticity theory by method of integral equations* // Investigation on plates and shells calculation. Rostov-on Don: RISI. 1986. P. 102-107. (in Russian).

[89] Katsevich A.Ya. *Numerical methods for solution of certain three-dimensional problems of elasticity theory* // Author's abstract of Ph.D. thesis. Rostov-on Don: 1987. 20 p. (in Russian).

[90] Ketov A.V. *Development of non-uniform load distribution calculation method in cylindrical Novikov gearing with account of its manufacturing errors* // Author's abstract of Ph.D. thesis. Kurgan: 1986. 16 p. (in Russian).

[91] Ketov A.V. *Account of manufacturing errors during calculation of non-uniform load distribution in cylindrical Novikov gearing*. Kurgan: 1986. 19 p. Dep. in VNIITEMR 07.01.86. N 3 msh - 86. (in Russian).

[92] Kirichenko A.F., Pavlov A.I. *Mathematical model of cylindrical Novikov gearing dynamics // Increasing the technical level, development of calculation and design methods of gearing, gearboxes and their parts*. Reports theses of republic scientific-technical conference. Kharkov: 1974. P. 82-85. (in Russian).

[93] Kovalenko G.D., Dementyev V.I., Volchinsky A.G. *Testing bevel gearing of a new tram KTM-5M traction gearbox* // Testing running gear of tram KTM-5M. Kharkov: 1969. P. 4-10 (in Russian).

[94] Kovalenko G.D. *Manufacturing enhancement of traction gearboxes of high-speed tram carriages RVZ-7 // Reliability increasing and metal capacity reduction of general-purpose gearing and gearboxes*. Reports theses of republic scientific-technical conference. Kharkov: 1983. P. 131-132 (in Russian).

[95] Kovalsky B.S. *Contact problem in engineering practice* // Izvestiya vuzov. Mashinostroenie. 1960. N 6. P. 30-36 (in Russian).

[96] Kovalsky B.S. *Calculation of parts for local compression*. Kharkov: KhVKIU. 1967. 233 p. (in Russian).

[97] *Conical hob: Inventor's certificate of USSR 1060361. B.I. 1983. N 46. (in Russian).*

[98] Korotkin V.I. *Cylindrical Novikov gearing adaptability and control of its gear teeth cutting* // Vestnik mashinostroeniya. 2005. N 6. P. 22-26 (in Russian).

[99] Korotkin V.I., Akhtyrets G.P. *On application of numerical method of solving contact problem for Novikov gearing* // Izvestiya SKNTs VSh. Estestvennye nauki. Rostov-on-Don: RSU Publishers. 1979. N 3. P. 24-27 (in Russian).

[100] Korotkin V.I., Akhtyrets G.P., Kharitonov Yu.D. *Non-uniform bending stress distribution along cylindrical Novikov gearing teeth length* // Vestnik mashinostroeniya. 1994. N 2. P. 6-8 (in Russian).

[101] Korotkin V.I. *Influence of center distance errors in cylindrical Novikov gearing with two lines of action* // Vestnik mashinostroeniya. 1977. N 4. P. 22-25 (in Russian).

[102] Korotkin V.I. *Influence of toothing width on load-bearing capacity of TLA Novikov gearing // Modern problems of continuum mechanics*. Proceedings of VII International conference. Vol.2. Rostov-on-Don: Publishers of "OOO "TsVVR". 2002. P. 105-109 (in Russian).

[103] Korotkin V.I. *Geometrical aspect of cylindrical Novikov gearing* // Vestnik mashinostroeniya. 2004. N10. P. 11-14 (in Russian).

[104] Korotkin V.I., Dorozhkin V.N. *Limiting geometrical factors for Novikov gearing* // Continuum mechanics. Rostov-on-Don: RSU Publishers. 1982. P. 34-42 (in Russian).

[105] Korotkin V.I., Dorozhkin V.N. *On some geometrical peculiarities of Novikov gearing* // *Problems of gearing quality and manufacturing technique effectiveness*: Reports theses of scientific-technical conference. Omsk: 1979. P. 50-53 (in Russian).

[106] Korotkin V.I. *Novikov gearing as a reserve of increasing the load-bearing capacity of general-purpose gearboxes* / Proceedings of the 36th conference on machine parts. Vol.1. Brno: 1995. P. 187-190 (in Russian).

[107] Korotkin V.I. *On calculation of hardened gears of Novikov gearing for depth contact endurance* // Izvestiya vuzov. Mashinostroenie. N 10. P. 44-48 (in Russian).

[108] Korotkin V.I., Minchenko M.E., Dorozhkin V.N. *Novikov gearing with straightened gear teeth for final gearing of a tractor* // *Increasing the gearing reliability and lifetime in tractor and agricultural machinery*. Reports theses of scientific-technical meeting. Kharkov: 1979. P. 22-25 (in Russian).

[109] Korotkin V.I. *Stresses in teeth of Novikov gearing and ways of their decreasing* // Theory and practice of gearing. Izhevsk: 1998. P. 96-100.

[110] Korotkin V.I. *On account of edge effects during analysis of Novikov gearing for contact endurance* // Vestnik mashinostroeniya. 1997. N 6. P. 8-11 (in Russian).

[111] Korotkin V.I. *On choice of safety factors during cylindrical Novikov gearing analysis* // Vestnik mashinostroeniya. 2002. N 6. P.13-16 (in Russian).

[112] Korotkin V.I., Onishkov N.P. *Novikov gearing – achieved results and ways of further development / Modern problems of science of machines and high technology*. Proceedings of International scientific-technical conference. Vol.1. DGTU Publishers. Rostov-on-Don: 2005. Vol. 1. P. 23-30 (in Russian).

[113] Korotkin V.I., Onishkov N.P. *On question of load-bearing capacity prediction in hardened Novikov gearing* // *Modern problems of continuum mechanics*. Proceedings of the V International conference. Vol.1 Rostov-on-Don: Publishers of SKNTs VSh. 2000. P. 106-110 (in Russian).

[114] Korotkin V.I., Onishkov N.P. *On choice of rational parameters of hardened layer for chemically heat treated parts* // *Modern problems of continuum mechanics*. Proceedings of the VII International conference. Vol.1. Rostov-on-Don: Publishers of "OOO "TsVVR". 2002. P. 150-153 (in Russian).

[115] Korotkin V.I. *On minimum teeth number of double-point contact engagement in cylindrical Novikov gearing* // Vestnik mashinostroeniya. 1977. N 3. P. 11-13 (in Russian).

[116] Korotkin V.I., Onishkov N.P. *On depth contact strength of surface-hardened Novikov gearing* // Problemy mashinostroeniya i nadezhnosti mashin. 2002. N 1. P. 42-46 (in Russian).

[117] Korotkin V.I. *On parameters of engagement quality of cylindrical Novikov gearing* // Vestnik mashinostroeniya. 2004. N 11. P. 3-6 (in Russian).

[118] Korotkin V.I. *Increasing load-bearing capacity in cylindrical Novikov gearing by longitudinal modification of tooth flanks* // Vestnik mashinostroeniya. 2003. N 5. P. 16-22 (in Russian).

[119] Korotkin V.I., Pozharsky D.A. *Die indentation in elastic three-dimensional wedge as a model of gearing tooth surfaces contact interaction* // Problemy mashinostroeniya i nadezhnosti mashin. 1996. N 3. P. 107-113 (in Russian).

[120] Korotkin V.I., *Pozharsky D.A. Friction influence on Novikov gearing contact strength and adaptability* // Problemy mashinostroeniya i nadezhnosti mashin. 2000. N 3. P. 58-65 (in Russian).

[121] Korotkin V.I. *Longitudinal modification of tooth flanks as a reserve of load-bearing capacity increasing in cylindrical Novikov gearing // Modern problems of continuum mechanics.* Proceedings of the VII International conference. Vol.1. Rostov-on-Don: Novaya kniga. 2003. P. 87-90 (in Russian).

[122] Korotkin V.I. *Straight-tooth bevel gearing with point contact, obtained by a special machine-tool setting* // Author's abstract of Ph.D. thesis. Rostov-on-Don: 1965. 22 p. (in Russian).

[123] Korotkin V.I. *Design loads in cylindrical Novikov gearing* // Vestnik mashinostroeniya. 2005. N 8. P. 11-15 (in Russian).

[124] Korotkin V.I. *Inverse problem solution for cylindrical Novikov gearing with straightened teeth and axial shift* // Continuum mechanics. Rostov-on-Don: RSU Publishers. 1988. P. 91-102 (in Russian).

[125] Korotkin V.I., Roslivker E.G. *Technique of geometrical and control parameters calculation in cylindrical Novikov gearing with account of peculiarities of cutting the basic rack profile configuration.* Rostov-on-Don: 1986. 16 p. Dep. in TsNIITEITYaZhMASh 18.09.86. N 1746-tm. (in Russian).

[126] Korotkin V.I. *Synthesis of new types of cylindrical gearing with convex-concave point contact // Izv. SKNTsVSh.* Tekhnicheskie nauki. Rostov-on-Don: RSU Publishers. 1979. N 4. P. 62-66 (in Russian).

[127] Korotkin V.I. *Degree of contact pattern localization in straight-tooth bevel gears* // Stanki i instrument. 1965. N 3. P. 18-20 (in Russian).

[128] Korotkin V.I., Kharitonov Yu.D. *Blocking contours for bevel Novikov gearing* // Vestnik mashinostroeniya. 2000. N 5. P. 26-29 (in Russian).

[129] Korotkin V.I., Kharitonov Yu.D. *Novikov gearing.* Rostov-on-Don: RSU Publishers. 1991. 208 p. (in Russian).

[130] Korotkin V.I., Kharitonov Yu.D. *Bevel Novikov gearing with modified tooth surfaces // Modern problems of continuum mechanics.* Proceedings of the VI International conference. Vol. 2. Rostov-on-Don: Publishers of SKNTs VSh. 2001. P. 90-93 (in Russian).

[131] Korotkin V.I., Kharitonov Yu.D. *Longitudinal tooth modification in order to increase load-bearing capacity of bevel Novikov gearing* // Problemy mashinostroeniya i nadezhnosti mashin. 1998. N 2. P. 72-76 (in Russian).

[132] Korotkin V.I., Kharitonov Yu.D., Fomenko V.E. *Calculation of bending and contact stresses of bevel Novikov gearing teeth* // Vestnik mashinostroeniya. 1998. N 3. P. 6-8 (in Russian).

[133] Korotkin V.I. *Cylindrical Novikov gearing with reduced axial force in engagement* // Stanki i instrument. 1980. N 4. P. 38 (in Russian).

[134] Korotkin V.I., Yavriyants Z.A. *On determining the value of interacting of active parts of tooth flanks in cylindrical Novikov gearing.* Reports theses of scientific-technical

conference “Problems of gearing and gearbox industry”. Kharkov: 1993. P. 40 (in Russian).

[135] Korotkin V.I., Yavriyants Z.A. *Determination of real profile of Novikov gears teeth and its active parts.* Rostov-on-Don: 1993. 25 p. Dep. In VINITI 23.07.93. N 2099-V93. (in Russian).

[136] Kotov O.K. *Surface hardening of machine parts by chemically heat methods.* Moscow: Mashinostroenie. 1969. 344 p. (in Russian).

[137] Kouba Yu.F. *Implementation experience of gearing with double-point contact engagement* // Novikov gearing. Issue 1. Moscow: TsINTIAM. 1964. P. 7-17 (in Russian).

[138] Kouba Yu.F. *Application experience of new progressive manufacturing process of tooth cutting of bevel pinions with double-point contact Novikov gearing // Results of investigation and practical application of M.L. Novikov gearing.* Reports theses of republic scientific-technical conference. Kharkov: 1971. P. 146-149 (in Russian).

[139] Krasnoshchekov N.N., Fedyakin R.V., Chesnokov V.D. *Theory of Novikov gearing.* Moscow: Nauka. 1976. 174 p. (in Russian).

[140] Kudryavtsev V.N., Derzhavets Yu.A., Glukharev E.G. *Design and analysis of gearboxes.* Reference book. Leningrad: Mashinostroenie. 1971. 328 p. (in Russian).

[141] Kudryavtsev V.N. *Simplified calculations of gearing.* Leningrad: Mashinostroenie. 1967. 112 p. (in Russian).

[142] Kurgansky V.I. *Experience of high-power high-speed Novikov gearing operation // Increasing the gearing reliability and lifetime in tractor and agricultural machinery.* Reports theses of scientific-technical meeting. Kharkov: 1979. P. 181-182 (in Russian).

[143] Kuchma V.Ya. *Investigation of bevel Novikov gearing accuracy // Reducing of gearing and gearboxes metal intensity.* Reports theses. Sverdlovsk: 1984. P. 35 (in Russian).

[144] Lashnev S.I., Yulikov M.I. *Calculation and design of metal-cutting tools by means of computer.* Moscow: Mashinostroenie. 1975. 392 p. (in Russian).

[145] Lingaya, Ramachandra. *Coefficient of adjacency in Wildhaber-Novikov gearing with tooth profiles, drawn along a circumference arc* // Konstruirovanie i tekhnologiya mashinostroeniya. 1981. N 1. P. 134-140 (in Russian).

[146] Litvin F.L. *Theory of gearing. 2nd edition, worked over and added.* Moscow: Nauka. 1968. 584 p. (in Russian).

[147] Lifshits R.I., Ivanov B.D. *Peculiarities of bevel Novikov gearing analysis // Novikov gearing.* Proceedings of NIITM. Issue X. Rostov-on-Don: 1964. P. 120-134 (in Russian).

[148] Lopato G.A., Kabatov N.F., Segal M.G. *Spiral bevel and hypoid gearing.* Reference book. Moscow: Mashinostroenie. 1977. 423 p. (in Russian).

[149] Lopato G.A. *Tooth number standardization of power spiral bevel pairs* // Vestnik mashinostroeniya. 1984. N 4. P. 19-20 (in Russian).

[150] Lyukshin V.S. *Theory of helical surfaces in design of cutting tools.* Moscow: Mashinostroenie. 1968. 371 p. (in Russian).

[151] *Machines and test-rigs for parts testing* / Ed. by D.N. Reshetov. Moscow: Mashinostroenie. 1979. 343 p. (in Russian).

[152] *Technique for determination of coefficient, taking into account non-uniform distribution of bending stresses in teeth of cylindrical TLA Novikov gearing* / Akhtyrets G.P.,

Korotkin V.I., Romanov R.A., Kharitonov Yu.D. // Vestnik mashinostroeniya. 1995. N 6. P. 18-20 (in Russian).

[153] Minchenko S.I. *On question of bevel pairs geometry* // Design and implementation experience of M.L. Novikov gearing. Moscow: GOSINTI. 1961. P. 80-95 (in Russian).

[154] Miroshnikov V.A., Pavlenko A.V. *Theoretical investigation of design geometrical parameters influence on fracture strength of bevel gears circular teeth // Results of investigation and practical application of M.L. Novikov gearing.* Reports theses of republic scientific-technical conference. Kharkov: 1971. P. 63-66 (in Russian).

[155] Mochalov V.L., Vasin G.M. *Determination of backlash between gear teeth in non-zero Novikov gearing* // Mechanical gearing. Izhevsk: "Udmurtiya" Publishers. 1972. P. 173-180 (in Russian).

[156] *Load-bearing capacity of Novikov gearing with new basic rack profile type, combining involute and circumference* // Ekspress-informatsiya. Detali mashin. 1987. N 16. P. 1-14 (in Russian).

[157] Nikitin B.G. *On one case of application of various-profiles gearing* // Mechanical gearing. Izhevsk: "Udmurtiya" Publishers. 1972. P. 181-185 (in Russian).

[158] Novikov M.L. Gearing with new type of engagement. Moscow: ZAFEA. 1958. 186 p. (in Russian).

[159] Ognev M.E. *Test results of non-zero cylindrical TLA Novikov gearing* // Vestnik mashinostroeniya. 1980. N 8. P. 16-17 (in Russian).

[160] Onishkov N.P. *Local stress condition and evaluation of depth contact strength of surface hardened Novikov gearing* // Ph.D. thesis. Kharkov: 1991. 164 p. (in Russian).

[161] Onishkov N.P., Onishkov I.N. *Prediction of diffusion layer characteristics in chemically heat hardened parts. // Modern problems of continuum mechanics.* Proceedings of the VII International conference. Vol.2. Rostov-on-Don: Publishers of "OOO "TsVVR". 2002. P. 125-129 (in Russian).

[162] Onishkov N.P. *Depth contact strength evaluation in surface hardened parts. // Machine parts.* Republic interdepartmental scientific-technical digest. Issue 51, Kiev: Tekhnika. 1991. P. 95-101 (in Russian).

[163] *On compliance of gear teeth in cylindrical Novikov gearing* / Akhtyrets G.P., Korotkin V.I., Pavlenko A.V., Fomenko V.E. // Izvestiya vuzov. Mashinostroenie. 1989. N 8. P. 27-31 (in Russian).

[164] *Stress determination in risky sections of complex shape parts.* / Verkhovsky A.V., Andronov V.I., Ionov V.A., Lupanova O.K., Chervikov V.I. Moscow: Mashgiz. 1958. 147 p. (in Russian).

[165] Experience of *Novikov gearing application in power transmissions // Ekspress-informatsiya.* Detali mashin. 1987. N 17. P. 18-24 (in Russian).

[166] Orlov A.V., Chermensky O.N., Nesterov V.M. *Testing of structural materials for contact fatigue.* Moscow: Mashinostroenie. 1980. 110 p. (in Russian).

[167] *Sharpened cutter heads for spiral bevel gearing.* Minsk: INTIP. 1964. 4 p. (in Russian).

[168] Pavlenko A.V., Baindurov V.S. *Bending stress distribution along transmission curve profile in lateral section of tooth model // Results of investigation and practical application of M.L. Novikov gearing.* Reports theses of republic scientific-technical conference. Kharkov: 1971. P. 67-69 (in Russian).

[169] Pavlenko A.V., Klimash O.S. *On Novikov gearing tooth shape change in the process of run-in // Theory and geometry of three-dimensional engagement.* Reports theses of the III USSR symposium. Kurgan: 1979. P. 80-82 (in Russian).
[170] Pavlenko A.V., Toporsky V.K. *Investigation of some geometrical properties of bevel Novikov gearing* // Results of investigation and practical application of M.L. Novikov gearing. Reports theses of republic scientific-technical conference. Kharkov: 1971. P. 3-7 (in Russian).
[171] Pavlenko A.V., Ukazov V.P. *Investigation of stresses in contact area of Novikov gearing teeth* // Nadezhnost i kachestvo zubchatykh peredach. Moscow: NIINFORMTYaZhMASh. 1967. N 18-67-66. 5 p. (in Russian).
[172] Pavlenko A.V., Fedyakin R.V., Chesnokov V.A. *Novikov gearing.* Kiev: Tekhnika. 1978. 144 p. (in Russian).
[173] Pavlenko A.V., Fedyakin R.V., Chesnokov V.A. *Experience of Novikov gearing industrial application and batch production.* Kiev: Obshchestvo "Znanie". 1982. 24 p. (in Russian).
[174] Paulinsh K.K. *Geometry-setting calculations of bevel Novikov gearing with modification near faces // Modern physical mechanical machining and control methods in precise machine-building.* Interinstitutional scientific-technical digest. Riga: 1978. P. 70-82 (in Russian).
[175] Paulinsh K.K. *Design-technological gearing system "Riga". Riga: Latv.* NIINTI. 1988. 4 p. (in Russian).
[176] *Novikov gearing with teeth surface hardness HB≥ 350. Strength analysis.* Methodical recommendations MR 221-86. Moscow: VNIINMASh. 1987. 86 p. (in Russian).
[177] Cylindrical TLA *Novikov gearing with teeth hardness 35 HRC and more. Basic rack profile.* Guidance on standardization RD2N24-11-88. Moscow: VNIITEMR. 1988. 34 p. (in Russian).
[178] Pisarenko G.S., Lebedev A.A. *Deforming and materials strength in complex stressed state.* Kiev: Naukova Dumka. 1976. 415 p. (in Russian).
[179] Plotnikov L.P. *Influence of basic rack profile and center distance deviations of cylindrical Novikov gearing on engagement geometry* // Nadezhnost i kachestvo zubchatykh peredach. Moscow: NIINFORMTYaZhMASh. 1967. N 18-67-75. 10 p. (in Russian).
[180] Plotnikov N.D. *Simulation on electronic digital computer of technological processes of hypoid and bevel teeth generation and conjugation control.* Ph.D. thesis. Saratov: 1973. 174 p. (in Russian).
[181] *Application of finite element method for determination of gears' stressed state* / Zhuravlev G.A., Zaguskin V.L., Iofis R.B., Reznitsky L.I. // Izvestiya SKNTs VSh. Estestvennye nauki. Rostov-on-Don: RSU Publishers. 1974. N 4. P. 34-37. (in Russian).
[182] *Three-dimensional gearing. Patent USA 3533300. B.I. 1970. N 2.*
[183] *Raiko M.V. Gearing lubrication.* Kiev: Tekhnika. 1970. 196 p. (in Russian).
[184] Raskin L.N. *Technical level increasing in mine winders // Reliability increasing and metal intensity reducing of general-purpose gearing and gearboxes.* Reports theses of republic scientific-technical conference. Kharkov: 1983. P. 6-7 (in Russian).

[185] *Load distribution between contact areas in real cylindrical Novikov gearing* / Akhtyrets G.P., Korotkin V.I., Pavlenko A.V., Fomenko V.E. // Izvestiya vuzov. Mashinostroenie. 1990. N 5. P. 12-16. (in Russian).

[186] *Strength analysis in machine-building.* Vol.1 / Ed. by S.D. Ponomarev. Moscow: Mashgiz. 1956. 884 p. (in Russian).

[187] *Gearboxes and gear-motors.* Catalogue. Part 1. Moscow: VNIITEMR. 1987. 66 p. (in Russian).

[188] Reizin A.G., Pavlenko A.V. *Comparative evaluation of influence of axes misalignment and helix angle error on kinematic accuracy in Novikov gearing // Results of investigation and practical application of M.L. Novikov gearing.* Reports theses of republic scientific-technical conference. Kharkov: 1971. P. 54-57 (in Russian).

[189] Reshetov D.N. *Machine parts.* Moscow: Mashinostroenie. 1975. 518 p. (in Russian).

[190] Romalis M.M. *On optimization method applied to engagement synthesis // Theory and geometry of three-dimensional engagement.* Reports theses of the III USSR symposium. Kurgan: 1979. P. 43-45 (in Russian).

[191] Romanov R.A., Kharitonov Yu.D. *Investigation of active tooth flanks of bevel Novikov gears with cycloid teeth // Experience of Novikov gearing investigation, design, manufacture and operation.* Reports theses of interrepublic scientific-technical conference. Riga: 1989. P. 23-24 (in Russian).

[192] Roslivker E.G., Dreizen I.S. *New gearbox* // Technical economical information bulletin of Rostov economical region SNH. Rostov-on-Don: 1958. N 4. P. 15-16 (in Russian).

[193] Roslivker E.G. *Kinematic investigation of gearing // Questions of reliability increasing in fusion cranes and dredging machines.* Proceedings of GIIVT. Issue 177. Part 1. Gorky: 1980. P. 158-175 (in Russian).

[194] Roslivker E.G., Kopasenko V.V. *Determination of normal stresses near tooth root of Novikov gearing.* Proceedings of NIITM. Issue X. Rostov-on-Don: 1964. P. 161-181 (in Russian).

[195] Roslivker E.G., Korotkin V.I. *Cutting straight-tooth bevel gears with barrel-shaped teeth* // Stanki i instrument. 1963. N 8. P. 14-17 (in Russian).

[196] Roslivker E.G. *Tooth strength and stiffness of M.L. Novikov gearing // Ship hulls strength and machine parts reliability.* Proceedings of GIIVT. Issue 138. Gorky: 1975. P. 50-96 (in Russian).

[197] Roslivker E.G. *Accuracy and control of cylindrical Novikov gearing // M.L. Novikov gearing. Proceedings of NIITM.* Issue X. Rostov-on-Don: 1964. P. 139-160 (in Russian).

[198] Rudenko S.P. *Resistance of case-hardened gears to contact fatigue. // Vestnik mashinostroeniya.* 1999. N 4. P. 13–15 (in Russian).

[199] Ryzhik M.A. *On question of Novikov gears contact strength analysis// M.L. Novikov gearing. Proceedings of NIITM.* Issue X. Rostov-on-Don: 1964. P. 182-207 (in Russian).

[200] Sakhonko I.M. *Contact endurance of hardened steel depending on geometrical parameters of touching bodies.* // Contact strength of machine-building materials. Moscow: Nauka. 1964. P. 52-59 (in Russian).

[201] Sevryuk V.N. *Bevel Novikov gearing.* Lvov: Publishers of LGU. 1968. 126 p. (in Russian).

[202] Sevryuk V.N. *Generalized analytical theory of circular surfaces and its application in gearing design* // Doct. Tech. Sc. thesis. Lvov: 1972. 246 p. (in Russian).

[203] Sevryuk V.N. *Theory of circular helical surfaces in Novikov gearing design.* Kharkov: Publishers of KhGU. 1972. 168 p. (in Russian).

[204] Segal M.G. *On contact localization in bevel and hypoid gearing // Three-dimensional and hypoid gearing.* Proceedings of NPI. Novocherkassk: 1970. Vol. 213. P. 124-135 (in Russian).

[205] Sedokov L.M., Martynenko A.G., Simonenko G.A. *Radial compression as a mechanical testing method.* // Zavodskaya laboratoriya. 1977. N 1. P. 98-100 (in Russian).

[206] Serensen S.V., Kogaev V.P., Shneiderovich R.M. *Bearing capacity and machine parts calculation.* Moscow: Mashinostroenie. 1975. 488 p. (in Russian).

[207] Silich A.A. *Development of geometrical theory of Novikov gearing design and process of gears tooth generation* // Author's abstract of Doct. Tech. Sc. thesis. Izhevsk: 1999. 31 p. (in Russian).

[208] Snesarev G.A. *Reserves of general gearbox industry* // Vestnik mashinostroeniya. 1990. N 8. P. 30-36. (in Russian).

[209] *Development of turbine compressor gearbox by Novikov gearing application* / Kirichenko A.F., Pavlenko A.V., Kuzovkov B.P., Maistrenko N.G. // Increasing of technical level, development of calculation and design methods of gearing, gearboxes and their parts. Reports theses of republic scientific-technical conference. Kharkov: 1974. P. 24-27. (in Russian).

[210] *Resistance to contact fatigue of heavy-loaded gears made of steel 20HN3A, hardened by chemically heat treatment.* / Shapochkin V.I., Zaitseva I.D., Burenkova O.S., Chebotarev F.M. // MiTOM. 1987. N 5. P. 12–15 (in Russian).

[211] Spektor A.G., Zelbot S.A., Kiseleva B.M. *Structure and properties of bearing steel.* Moscow: Metallurgiya. 1980. 258 p. (in Russian).

[212] Spektor A.G. *Internal geometry optimization and calculation of contact interaction characteristics of cylindrical roller bearings.* / Proceedings of NPO VNIIPP. Moscow: 1987. N 3. P. 60–74 (in Russian).

[213] A *method of TLA Novikov gearing manufacture: Inventor's certificate of USSR 1489068. B.I. 1989. N 23 (in Russian).*

[214] *A method of cutting bevel gears with curvilinear teeth: Inventor's certificate of USSR 290801. B.I. 1971. N 3 (in Russian).*

[215] Taits B.A. *Accuracy and control of gearing*. Moscow: Mashinostroenie. 1972. 367 p. (in Russian).

[216] Tellian T. *Influence of material properties and work conditions on rolling bearings durability* // Problemy treniya i smazki. 1988. N 4. P. 1–7 (in Russian).

[217] Tellian T. *Common model of predicting the rolling contact durability* // Problemy treniya i smazki. 1982. N 3. P. 35–48 (in Russian).

[218] Tesker E.K. *On calculation of surface-hardened gears for contact strength under action of maximum load* // Vestnik mashinostroeniya. 1987. N 3. P. 14-19 (in Russian).

[219] Timoshenko S.P., Gudyer J. *Elasticity theory.* / Ed. by G.S. Shapiro. Moscow: Nauka. 1975. 576 p. (in Russian).

[220] Todorov V.S. *Investigation and calculation of bevel Novikov gearing as applied for drives of mining machines* // Author's abstract of Ph.D. thesis. Dnepropetrovsk: 1970. 24 p. (in Russian).

[221] Trubin G.K. *Contact fatigue of gears materials*. Moscow: Mashgiz. 1962. 404 p. (in Russian).

[222] Turunovsky V.M. *Novikov gearing application in centrifugal compressors multipliers* // Vestnik mashinostroeniya. 1980. N 10. P. 16-18 (in Russian).

[223] *Unification of some parameters of double-point contact Novikov gearing in general-purpose gearboxes* / Kanaev A.S., Vasin G.M., Mochalov V.L., Ognev M.E. // Mechanical gearing. Izhevsk: "Udmurtiya" Publishers. 1974. P. 44-49 (in Russian).

[224] Fedyakin R.V. On M.L. *Novikov gearing // Calculation, design and investigation of gearing*. Proceedings of the conference. Part 1. Odessa: 1958. P. 129-139 (in Russian).

[225] Foskamp A.P. *Change in material under action of contact load during rolling*. // Problemy treniya i smazki. 1985. N 3. P. 35–43 (in Russian).

[226] Fudzita K., Iosida A. *Influence of case-hardened layer depth and relative curvature radius on durability at contact fatigue of case-hardened roller made of chromium-molybdenum steel* // Konstruirovanie i tekhnologiya mashinostroeniya. 1981. N 2. P. 115-124 (in Russian).

[227] Kharitonov Yu.D. *Kinematics analysis and synchronization of contacting along several lines of action of approximate bevel Novikov gearing*. Rostov-on-Don: 1986. 15 p. Dep. In TsNIITEITYaZhMASh 18.09.86. N 1749. (in Russian).

[228] Kharitonov Yu.D., Korotkin V.I. *Calculation of dynamic load coefficient for cylindrical Novikov gearing* // Vestnik mashinostroeniya. 1993. N 9. P. 25-26 (in Russian).

[229] Kharitonov Yu.D., Korotkin V.I. *Synthesis of bevel Novikov gearing with given longitudinal modification of tooth surfaces* // STIN. 2001. N 5. P. 14-17 (in Russian).

[230] Kharitonov Yu.D. *Tooth shape modification of bevel Novikov gearing for improvement of contacting conditions* // Ph.D. thesis. Kharkov: 1985. 270 p. (in Russian).

[231] Kharitonov Yu.D., Pavlenko A.V. *Mutual influence of lines of action in the gearing* // Theory of real gearing. Reports theses of the IV USSR symposium. Part 2. Kurgan: 1988. P. 53-54 (in Russian).

[232] Kharitonov Yu.D., Pavlenko A.V., Onishkova T.V. *Influence of teeth-cutting technological parameters at intersected and skewed axes of machine-tool engagement on bevel TLA Novikov gearing kinematics*. Rostov-on-Don. 1987. 31 p. Dep. In TsNIITEITYaZhMASh 19.06.87. N 1944-tm. (in Russian).

[233] Kharitonov Yu.D., Pavlenko A.V. *Theoretical fundamentals of tooth flanks development in bevel Novikov gearing with improved contacting conditions and longitudinal shape* // Theory of mechanisms and machines. Digest. Issue 43. Kharkov: "Vishcha shkola". 1987. P. 88-95 (in Russian).

[234] Kharitonov Yu.D., Safronova A.Yu. *Investigation of bevel Novikov gearing surface undercut by a method of mathematical simulation of teeth machining process // Dynamics, strength and reliability of agricultural machines*. Interinstitutional book of scientific proceedings. Rostov-on-Don: DGTU. 1996. P. 164-168 (in Russian).

[235] Kharitonov Yu.D. *The cutting method of generated bevel gears with circular teeth of equal height* // Technological methods of providing gearing quality. Reports theses of USSR Scientific-technical conference. Part 2. Moscow: 1981. P. 187-190 (in Russian).

[236] Tsepkov A.V. *Form-relieved tools profiling*. Moscow: Mashinostroenie. 1979. 150 p. (in Russian).

[237] *Cylindrical Novikov gearing: Inventor's certificate of USSR 544793. B.I. 1977. N 4*. (in Russian).

[238] *Cylindrical Novikov gearing: Inventor's certificate of USSR 580391. B.I. 1977. N 42*. (in Russian).

[239] *Cylindrical Novikov gearing* / Korotkin V.I., Roslivker E.G., Pavlenko A.V., Veretennikov V.Ya., Kharitonov Yu.D. Kiev: Tekhnika. 1991. 152 p. (in Russian).

[240] Chasovnikov L.D. *Gearing (toothed and worm)*. 2nd edition, worked over and added. Moscow: Mashinostroenie. 1969. 486 p. (in Russian).

[241] Cherny B.A. *Optimal synthesis of bevel gears approximate engagements* // Ph.D. thesis. Leningrad: 1974. 153 p. (in Russian).

[242] Shashin M.Ya. *Technique of determination of mean probable cyclic durability values // Zavodskaya laboratoriya*. 1966. N 6. P. 207-210 (in Russian).

[243] Shvarts A.I. *Operation of Novikov gearing at high tangential speeds // Nadezhnost i kachestvo zubchatykh peredach*. Moscow: NIINFORMTYaZhMASh. 1967. N 18-67-51. 9 p. (in Russian).

[244] Sheveleva G.I. *Elastic contact displacements calculation on surfaces of limited dimension parts // Mashinovedenie*. 1984. N 4. P. 92-98 (in Russian).

[245] Sheveleva G.I. *Universal programs for calculation of engagement on computer // Theory and geometry of three-dimensional engagement*. Reports theses of the II symposium. Leningrad: LDNTP. 1973. P. 33 (in Russian).

[246] Shishov V.P., Kuchma V.Ya. *On question of tolerances to manufacturing and assembly of bevel Novikov gearing with circular teeth // Reliability increasing and metal capacity reducing of general-purpose gearing and gearboxes*. Reports theses of republic scientific-technical conference. Kharkov: 1983. P. 128-129 (in Russian).

[247] Shunaev B.K., Efimenko V.F. *Diagonal tooth milling of gears with barrel-shaped teeth // Stanki i instrument*. 1967. N 4. P. 15-18. (in Russian).

[248] Yakovlev A.S. *On determination of bending stress in cylindrical gearing teeth by boundary finite element method // Mashinovedenie*. 1982. N 2. P. 89-94. (in Russian).

[249] Yakovlev A.S. Correction and technological effectiveness of cylindrical Novikov gearing // Progressive processes of toothed and worm gearing machining. Erevan: 1972. P. 117-129 (in Russian).

[250] Yakovlev A.S. *Determination of bending stress in teeth of cylindrical Novikov gearing // Vestnik mashinostroeniya*. 1984. N 6. P. 18-20 (in Russian).

[251] *A criterion for contact fatigue of ion-nitrided gear*. / Wu Y., Ma B., He J., Luo B. // Wear. 1989. N 1. P. 13–21.

[252] Ariga Y., Nagata S. *Development of a new W-N gear with a basic rack of combined circular and involute profile* / International Symposium on Gearing and Power Transmissions. Tokyo 1981. P. 87-92.

[253] Ariga Y., Nagata S. *Load Capacity of a New W-N Gear with Basic Rack of Combined Circular and Involute Profile*. // Trans. ASME. J. Mech. Transmiss. and Autom. Des. 1985. N 14. P. 565-572.

[254] *Bending stresses in bevel gear teeth* // Gear engineering standard. Gleason works. Rochester. N.Y. USA. 1965. 28 p.

[255] Chen Chen-Wen. *The relationships between Load parameter* $(F_n P / m_n^2 E')$ *of W-N gears and its stiffness parameter respectively* / International Symposium on Gearing and Power Transmissions. Tokyo. 1981. P. 93-98.

[256] Chen Chen-Wen, Chen Rong-Zeng, Ye Rei-Da. *Experimental study of the helical gear couple with stepped-double-circular-arc H-2 type tooth profile/ International Symposium on Gearing and Power Transmissions.* Tokyo. 1981. P. 99-104.

[257] *Effect of case depth of fatigue strength/*Yoshida A., Fujita K., Kanehara T., Ota K. // Bull. JSME. 1986. N 247. S. 228–234.

[258] Elkholy A.H. *Optimum carburization in extremal spur gears.* // Wear.1984. N 1. S. 79–86.

[259] *Helical gearing: Patent USA 1601750. 1926.*

[260] Hua Jia-shou. *Experience of Adopting Dual Circular-arc Profile (W-N) Gear Ondorb Ship / 2 eme Cong. mond. Engren. Paris. 3-5 mars.* 1986. Textes Conf. Vol. 2. S.l.s.a. 107-113.

[261] Jaramillo T.J. *Deflections and Moments due to Concentrated Load on a Cantilever-Plate of Infinite Length//* Journal of Applied Mechanics / Trans. ASME. 1950. Vol. 72. P. 67-72, 342-343.

[262] *Multiple-contact type W-N Gear: Patent USA 4051745. 1976.*

[263] Niemann G. *Novikov-Verzahnung and andere Verzahnungen fur hohe Tragfehigkeit //* VDI-Berichte. Dusseldorf. VDI. Verlag. 1961. Nr. 47.

[264] *Novikov gearing: Patent USA 5022280. 1991.*

[265] *Surface durability pitting formulae for bevel gear teeth //* Gear engineering standard. Gleason works. Rochester. N.Y. USA. 1966. 28 p.

[266] *Tooth gearing analysis and surface durability of symmetrical conformal gears / Kasuya K.,t Nogami M., Matsunaga T., Watanabe K.* // Journal of Mechanical Design. 1981. Vol. 103. HI. P. 141-151.

[267] *Understanding tooth contact analysis.* Gleason Works. Rochester. N.Y. U.S.A. 1970. 7 p.

[268] Wolkenstein R. Die hartertiefe an Zahnradern. *Teil 1. Zahnflanke-Prinzip des Verfarens.* // Maschinenwelt-Elektrotechn. 1979. N 5. S. 116–122.

[269] Xiao-gang L., Qing G., Eryu S. *Initiation and propagation of case crushing cracks in rolling contact fatigue.//* Wear. 1988. N 1. P. 33–43.

INDEX

D

E

F

G

H

I

J

K

L

M

N

O

P

Q

R

S

T

U

V

W

Y